Collins

NEW MATHS FRAMEWORKING

Matches the revised KS3 Framework

Kevin Evans, Keith Gordon, Trevor Senior, Brian Speed

Contents

Introduction

Welcome to *New Maths Frameworking*!

New Maths Frameworking Year 9 Practice Book 2 has hundreds of levelled questions to help you practise Maths at Levels 5-7. The questions correspond to topics covered in Year 9 Pupil Book 2 giving you lots of extra practice.

These are the key features:

- **Colour-coded National Curriculum levels** for all the questions show you what level you are working at so you can easily track your progress and see how to get to the next level.

- **Functional Maths** is all about how people use Maths in everyday life. Look out for the Functional Maths icon (FM) which shows you when you are practising your Functional Maths skills.

Practice

1A Sequences

1 Find the next three terms in these sequences.

    a 4, 11, 18, 25 ...
    b 100, 92, 84, 76 ...
    c −25, −19, −13, −7 ...
    d 2.43, 2.54, 2.65, 2.76 ...
    e 3, 5, 8, 12, 17 ...
    f 1, 3, 7, 13, 21 ...

2 Write down the first four terms of the sequences with these nth terms.

    a $6n + 2$
    b $3n − 10$
    c $−2n + 3$
    d $\dfrac{n + 1}{2}$
    e $0.7n$
    f $2.3n − 0.4$

3 Find the nth term for each of these sequences.

    a 5, 10, 15, 20 ...
    b 2, 3, 4, 5 ...
    c 7, 10, 13, 16 ...
    d 8, 18, 28, 38 ...
    e −3, −6, −9, −12 ...
    f −10, −8, −6, −4 ...
    g 0.3, 0.6, 0.9, 1.2 ...
    h 6.6, 7.4, 8.2, 9 ...
    i $\frac{3}{2}, \frac{6}{3}, \frac{9}{4}, \frac{12}{5}$ ...

4 Find the nth term for each of these sequences.

    a 25, 24, 23, 22 ...
    b 12, 10, 8, 6 ...
    c 100, 95, 90, 85 ...
    d −4, −8, −12, −16 ...
    e −5, −8, −11, −14 ...
    f 7.2, 7, 6.8, 6.6 ...

5 These tables show Martha and Johan's bank balances during the first five weeks of 2003.

Johan's bank account					
Week	1	2	3	4	5
Balance, £	970	1035	1100	1165	1230

Martha's bank account					
Week	1	2	3	4	5
Balance, £	340	290	240	190	140

    a Write a formula to show the balance of Johan's bank account after n weeks of 2003.
    b Write a formula to show the balance of Martha's bank account after n weeks of 2003.
    c What will be the balance of **i** Johan's **ii** Martha's bank account after nine weeks of 2003?

Martha and Johan combine their bank balances.

    d Make a new table to show their combined bank balance during the first five weeks of 2003.
    e Write a formula to show their combined balance after n weeks of 2003.
    f Martha and Johan are saving for a holiday that costs £2200. When will they have saved enough?

1 This diagram shows a growing pattern of equilateral mosaic tiles.

Each new pattern is made by placing a tile alongside every edge of the previous pattern.

a Use triangular grid paper to draw the next two patterns.
b Copy and complete this table.

Pattern number, n	1	2	3	4
Number of tiles in pattern	1	4		

c Describe a rule for finding the number of tiles in a pattern.
d i Use your rule to calculate the number of tiles in the fifth pattern.
  ii Verify your rule works by drawing the fifth pattern.

2 This diagram shows how different numbers of flowerpots can be arranged in a row.

1 pot
1 arrangement

2 pots
2 possible arrangements

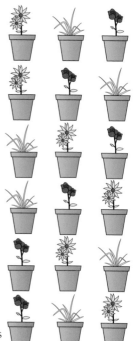

3 pots
6 possible arrangements

a Find the number of ways four flowerpots can be arranged in a row.

b Copy and complete this table.

Number of flower pots, n	1	2	3	4
Number of ways flower pots can be arranged in a row				

c Describe a rule for finding the number of ways n flowerpots can be arranged in a row.

d i Use your rule to calculate the number of ways five flower pots can be arranged.

ii Verify that your rule works by drawing a diagram.

Hint: You don't need to draw all of the arrangements.

Practice

1C Functions

1 Find the inverse of each function.

a $x \rightarrow 6x$ b $x \rightarrow x + 7$ c $x \rightarrow \frac{x}{3}$ d $y = x - 1.5$

2 Find the inverse of each function.

a $x \rightarrow 4x - 3$ b $x \rightarrow 10x + 2$ c $y = 2x - 4$ d $x \rightarrow \frac{x}{3} + 5$

3 Find the inverse of each function.

a $x \rightarrow 4(x + 2)$ b $y = 3(x - 5)$ c $x \rightarrow \frac{(x + 4)}{3}$

4 i Find the function for each mapping diagram.

ii Find the inverse of the function.

a $x \rightarrow$?
1 $\rightarrow$ 4
2 $\rightarrow$ 8
3 $\rightarrow$ 12
4 $\rightarrow$ 16

b $x \rightarrow$?
3 $\rightarrow$ 6
4 $\rightarrow$ 7
5 $\rightarrow$ 8
6 $\rightarrow$ 9

c $x \rightarrow$?
1 $\rightarrow$ 3
2 $\rightarrow$ 5
3 $\rightarrow$ 7
4 $\rightarrow$ 9

5 Which of these functions are i identities and ii self-inverse functions?

a $x \rightarrow 2x - 2$ b $x \rightarrow -10 - x$ c $x \rightarrow \frac{3x}{3}$ d $x \rightarrow \frac{2}{x}$

6 Each mapping diagram is of the inverse of a function.

i Find the inverse function.

ii Find the original function.

a ? $\leftarrow$ x
1 $\leftarrow$ 6
2 $\leftarrow$ 7
3 $\leftarrow$ 8
4 $\leftarrow$ 9

b ? $\leftarrow$ x
1 $\leftarrow$ 5
2 $\leftarrow$ 8
3 $\leftarrow$ 11
4 $\leftarrow$ 14

1 Ben and Gurjit left their homes at the same time and travelled to each other's home. This graph shows their journeys.

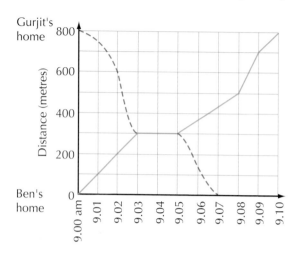

Give reasons for your answers to these questions.

a One person walked and ran, the other cycled. Which person cycled?

b Gurjit and Ben met each other.
  i When was this?
  ii How far were they from Gurjit's home?
  iii How long was their meeting?

c There is a hill between their two homes. Roughly how far from Ben's home is the top of the hill?

d When did Ben travel slowest?

e When did Gurjit travel fastest?

f Ben and Gurjit both started their return journeys at 9.30 am. Ben waited one minute at the top of the hill for Gurjit to arrive. They chatted for 2 minutes. Sketch their return journeys.

 2 Each of these candles is 25 cm tall and burns to the ground in 20 hours.

a

b

c
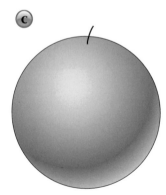

For each candle, sketch a graph showing its height over time.

FM **3** Match each graph with the correct description.

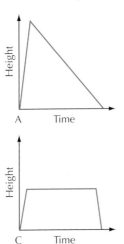

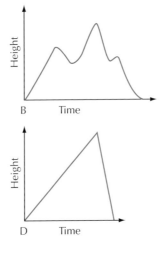

a A hovercraft journey.
b A rocket launch and parachute landing.
c A flight of a radio-controlled model aeroplane.
d A model hot air balloon that takes off and catches fire.

FM **4** Sketch a graph to show these.

a The height of a space shuttle that takes off, circles the Earth and lands.
b The weight of a woman 3 months before she is pregnant, during her pregnancy and 3 months after she gives birth.

Practice

1E Limits of sequences

1 **a** Calculate the first eight terms of this sequence:
  First term: 1
  Term-to-term rule: divide by 4 then add 6
b State the limit of the sequence.

2 **a** Calculate the first eight terms of this sequence:
  First term: 1
  Term-to-term rule: divide by 5 then add 10
b State the limit of the sequence.

3 **a** Copy and continue these calculations to give the first eight terms of a sequence.

$$1 = 1$$
$$1 + \frac{1}{2} = 1.5$$
$$1 + \frac{1}{2} + \frac{1}{4} =$$
$$1 + \frac{1}{2} + \frac{1}{4} + \frac{1}{8} =$$
$$1 + \frac{1}{2} + \frac{1}{4} + \frac{1}{8} + \frac{1}{16} =$$

b State the limit of the sequence, if there is one.

4 **a** Calculate, as decimals, the 10th, 100th and 1000th terms of the sequence $\frac{1}{2}, \frac{2}{3}, \frac{3}{4}, \frac{4}{5} \ldots$

b State the limit of this sequence.

5 **a** Calculate, as decimals, the first ten terms of the sequence whose nth term is $10 - 1/n$.

b Calculate, as decimals, the 100th and 1000th terms.

c State the limit of the sequence.

6 Find a sequence whose limit is 4. Write down its nth term.

Hint: Look at Question 3.

7 Use a spreadsheet to investigate the limit of the sequence produced by these calculations.

$$1 \qquad\qquad\qquad =$$

$$1 + \frac{1}{2} \qquad\qquad\qquad =$$

$$1 + \frac{1}{2} + \frac{1}{3} + \frac{1}{4} \qquad =$$

$$1 + \frac{1}{2} + \frac{1}{3} + \frac{1}{4} + \frac{1}{5} =$$

etc.

CHAPTER **2** Number **1**

Practice

2A Adding and subtracting fractions

Convert these fractions to equivalent fractions with the same denominator. Work out the answer. Cancel your answer and write as a mixed fraction, if necessary.

1 **a** $2\frac{1}{3} + 1\frac{2}{5}$ **b** $3\frac{1}{4} + 4\frac{2}{3}$ **c** $1\frac{5}{6} + 2\frac{2}{3}$

      d $3\frac{7}{9} + 6\frac{1}{6}$ **e** $7\frac{1}{6} + 5\frac{3}{4}$ **f** $5\frac{3}{8} + 4\frac{1}{10}$

2 **a** $3\frac{3}{4} - 1\frac{2}{3}$ **b** $5\frac{5}{6} - 2\frac{4}{5}$ **c** $4\frac{1}{8} - 1\frac{1}{3}$

      d $6\frac{5}{6} - 3\frac{1}{4}$ **e** $5\frac{5}{8} - 1\frac{9}{10}$ **f** $9\frac{1}{6} - 4\frac{7}{9}$

3 Find the LCM of each pair of numbers.

      a 6 and 15   **b** 16 and 20   **c** 12 and 18

4 Use your answers to Question 3 to help you to calculate these.

      a $\frac{5}{6} - \frac{7}{15}$ **b** $\frac{5}{16} + \frac{13}{20}$ **c** $\frac{13}{18} - \frac{7}{12}$

5 Calculate these.

   a $\frac{9}{10} + \frac{7}{15}$
   b $\frac{13}{14} - \frac{3}{8}$
   c $\frac{11}{12} - \frac{13}{24}$

6 Marion buys a tub containing $7\frac{7}{8}$ oz of peanuts and a packet containing $3\frac{5}{6}$ oz of peanuts.

   a What is the total weight of the peanuts she bought?
   b How much more does the tub contain than the packet?

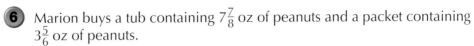

Practice

2B Multiplying and dividing fractions

Cancel before multiplying, if possible. Cancel your answer and write as a mixed number, if necessary.

1 Work out each of the following:

   a $\frac{4}{9} \times \frac{2}{5}$
   b $\frac{5}{8} \times \frac{2}{3}$
   c $\frac{8}{15} \times \frac{5}{7}$

   d $\frac{10}{21} \times \frac{7}{15}$
   e $\frac{9}{16} \times \frac{12}{27}$
   f $\frac{8}{9} \times \frac{6}{15} \times \frac{5}{12}$

2 Write as improper (top-heavy) fractions first.

   a $1\frac{2}{3} \times 1\frac{3}{4}$
   b $2\frac{4}{5} \times 1\frac{2}{7}$
   c $2\frac{5}{8} \times 3\frac{5}{9}$

   d $4\frac{4}{5} \times 3\frac{1}{3}$
   e $2\frac{7}{10} \times 2\frac{7}{9}$

3 Work out each of the following:

   a $\frac{5}{8} \div \frac{2}{3}$
   b $\frac{3}{5} \div \frac{7}{10}$
   c $\frac{15}{16} \div \frac{9}{10}$

   d $\frac{8}{9} \div \frac{10}{21}$
   e $\frac{16}{25} \div \frac{14}{15}$

4 Write as improper (top-heavy) fractions first.

   a $2\frac{2}{3} \div 1\frac{2}{5}$
   b $1\frac{1}{6} \div 2\frac{3}{4}$
   c $3\frac{5}{9} \div 3\frac{2}{3}$

   d $4\frac{1}{8} \div \frac{33}{40}$
   e $\frac{8}{15} \div 5\frac{3}{5}$

5 A bottle contains $7\frac{7}{8}$ ounces of perfume. A perfume spray contains $\frac{7}{10}$ ounces of perfume.

   a How many sprays can be filled from the bottle?
   b How much perfume is left over?

Practice

2C Percentages and compound interest

1 Write down the multiplier equivalent to each of these.

   a 15% increase
   b 2% decrease
   c 95% increase

   d 60% decrease
   e $12\frac{1}{2}$% increase
   f $4\frac{1}{4}$% decrease

   g 33.3% increase
   h 8.7% decrease
   i $17\frac{1}{2}$% increase

For each part of questions 2–4, write down:

i the multiplier
ii your calculation
iii the answer to the question.

2 How much would you have in the bank if you invested:

   a £500 at 4% interest for 3 years?

   b £8000 at $3\frac{1}{2}$% interest for 5 years?

   c £250 at 5.3% for 10 years?

3 In 1960, 8000 cars were abandoned in a city. Each year, 8% more cars were abandoned.

   a How many cars were abandoned in 1966?
   b Which was the first year when more than 20 000 cars were abandoned?
   c How many cars were abandoned in 2000?

4 A snowman has a volume of 600 litres. Each day, it loses 13% of its volume.

   a What will its volume be after 4 days?
   b How many days will it take for there to be less than 100 litres of snow left?
   c What will be its volume after 20 days?

Practice

2D Reverse percentages

Choose either the unitary method or the multiplier method to answer all the questions.

Set out your working using the following steps:

Unitary method	Multiplier method
Write down the percentage that the new amount represents (100% + % increase or 100% − % decrease)	Write down the multiplier (1 + the increase or 1 − the decrease) written as a decimal
Divide both sides by the percentage to get 1%	Divide the new amount by the multiplier to get the answer
Multiply both sides by 100 to get the answer	

1 The average water consumption of a household is 165 litres per day. This is an increase of 10% on last year. What was the consumption last year?

2 There are 56 spectators at a hockey match. This is 20% fewer than at the last match. How many spectators were at the last match?

3 From 1980 to 1998, the Royal Navy decreased its size by 17% to 28 500. How many people were in the Royal Navy in 1980?

4 What were the prices before VAT was added to these items?

a

£54 incl. VAT @ $17\frac{1}{2}\%$

b

£9.50 incl. VAT @ 17.5%

c

G A S B

Total £85.23

including
VAT @ 5%

Practice **2E Ratio**

1 Cancel these ratios to their simplest form.

   a 9 : 15   **b** 28 : 12   **c** 200 : 125 **d** 44 : 121   **e** 16 : 8 : 4

2 Write these ratios in the form 1 : n. Round your answers to 1 decimal place where necessary.

   a 3 : 18   **b** 10 : 25   **c** 7 : 4   **d** 40 : 50   **e** 27 : 60

3 Write these ratios in the form n : 1. Round your answers to 1 decimal place where necessary.

   a 28 : 4   **b** 42 : 8   **c** 200 : 110 **d** 2 : 5   **e** 10.5 : 2.4

4 The ratio of men to women attending a concert was 7 : 5. In total, 228 people attended.

   a How many of these were men?
   b How many of these were women?

5 Catriona spends her pocket money in this ratio:

    savings : magazines : entertainment = 5 : 3 : 2

How does she spend £4.50 of pocket money?

6 A compost mixture is made from peat and garden waste in the ratio 4 : 11.

   a How much peat does a 120-litre bag of compost mixture contain?
   b If a pile of compost mixture contains 48 litres of peat, how much garden waste does it contain?
   c How much compost mixture could be made from 60 litres of peat and 99 litres of garden waste?

7 Charities use part of their donations for running costs and the remainder for good causes. This table shows the ratios for three charities.

Charity	Running costs : Good causes
World Need	2 : 5
Pet Protection	13 : 30
Cancer Cure	8 : 21

Which charity gives the greatest proportion of donations to the good causes?

Hint: Convert the ratios to the form $1 : n$.

Practice

2F Numbers between 0 and 1

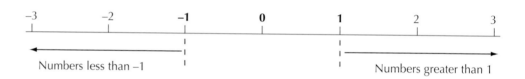

Use your calculator to help you to answer these questions.

 Describe the result of each calculation, **a** to **h**, by choosing one of these answers:

A number between −1 and 0
A number between 0 and 1
A number greater than 1
A number less than −1
A number less than 0
A number greater than 0
A number between −1 and 1
0
1
−1
Impossible
None of the above

a Squaring a number greater than 1
b Finding the square root of a number between 0 and 1
c Squaring −1
d Squaring a number less than −1
e Finding the square root of a number between −1 and 0
f Dividing −1 by a number less than −1
g Dividing 1 by a number between −1 and 0
h Dividing 0 by 0

2 **a** Describe the effect of multiplying any number by –1.
b Describe the effect of dividing any number by –1.

3 Describe the effect of dividing 1 by a number between –1 and 1.

4 **a**

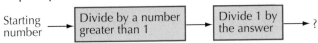

Work your way along the flow chart.
Describe the result using one of the answers shown in Question 1.

i Start with a number between 0 and 1.
ii Start with –1.
iii Start with a number between –1 and 1.

b Repeat part **a** with this flowchart.

Starting number ⟶ Divide by a number greater than 1 ⟶ Divide 1 by the answer ⟶ ?

Practice

2G Order of operations

Evaluate these. Show your working.

1 **a** $100 - 16 \div 2 \times 3$ **b** $48 \div (3 \times 2) + 2$ **c** $4^2 + 8 \times (20 - 17)$

d $(4^3 - 10) \div 2$ **e** $\dfrac{5 \times 8}{9 - 5}$ **f** $(3 + 2 \times 4)^2$

g $(4^2 - 12)^2$ **h** $(8 + 1)(8 - 1)$ **i** $2(3 \times 2)^2$

2 **a** $(10 + 2^3) \div (15 - 6)$ **b** $(2 \times 3^2)^2$ **c** $(5 + 2)(5 - 2)^2$

d $\dfrac{(5 - 1)^2}{(4 + 2)^2}$ **e** $(30 - 3^3)^2 \times (12 - 8)$

f $(6^2 + 3^2) \div (2^3 + 1)$ **g** $(9 - 1)^2(8 \div 4)^2$

3 **a** $(2 + 3)^3 - 2(3 + 4^2 \times 3)$ **b** $5 + 4(2^2 + 1) \div (10 - 2^3)$

4 Put brackets into these equations to make them true.

a $1 + 4 \times 2^2 + 3 - 6 = 23$ **b** $9 - 3^2 \div 3 \times 2 = 24$

5 Evaluate these. Cancel where possible. Show your working.

a $\dfrac{(3 \times 8)^2}{4 \times 6}$ **b** $\dfrac{7^2 \times 50 \times 16}{(10 \times 14)^2}$

7

1 Round these numbers to 1 significant figure.

    a 28   **b** 0.846   **c** 3521
    d 0.0665   **e** 962   **f** 12.29

2 Calculate these.

    a 400×60 **b** 2000×900 **c** 0.4×0.8
    d 0.02×0.7 **e** 0.04×0.09 **f** 300×0.2
    g 0.08×4000 **h** 60×0.9

3 Calculate these.

    a $800 \div 20$ **b** $12\,000 \div 300$ **c** $0.8 \div 0.4$
    d $0.12 \div 0.03$ **e** $6 \div 0.3$ **f** $40 \div 0.05$
    g $3000 \div 0.6$

4 Estimate answers to these by rounding the numbers to 1 significant figure first. Show your working.

    a 278×59 **b** $9483 \div 34$ **c** 0.358×166
    d $43 \div 0.537$ **e** 0.0366×0.0899 **f** $777 \div 0.241$
    g 3278^2 **h** $2.67 + 5.18 \times 9.01$ **i** $53 \times (0.672 - 0.306)$

5 This table shows the prices of some metals.

Metal	Price
Lead	£0.2412 per kg
Silver	£0.085 per g
Gold	£2575 per kg
Platinum	£11.63 per g

    a Estimate the cost of buying these. Show your working.
        i 2474 kg of lead   **ii** 0.2856 g of silver
        iii 0.036 kg of gold   **iv** 19.08 g of platinum
    b Estimate answers to these. Work in pounds. Show your working.
        i How much gold could be bought for £64 250?
        ii How much lead could be bought for £124.59?
        iii How much platinum could be bought for £289 320?
        iv How much silver could be bought for £0.94?

CHAPTER 3 Algebra 3

3A Equations, formulae and identities

Solve these equations. Show your working.

1   **a** $7(v + 3) = 5(v + 7)$ **b** $2(2d + 4) = 4(2d - 7)$
    c $3(x - 2) = -2(x + 19)$ **d** $2(5 - 2k) = 3(1 + 2k)$

2   **a** $3(m + 1) + 2(m + 2) = 52$ **b** $7(d + 2) - 2(2d + 1) = 21$
    c $2(6t + 1) - 4(t - 1) = 0$ **d** $4(2 - b) + 3(3 + 2b) = 33$

3   The time, T seconds, for a person to travel n floors in a lift is given by the formula $T = 10 + 7n$.

    a   How long does it take to travel:
       i   3 floors?   **ii**   9 floors?
    b   **i**   Calculate T when $n = 0$.
       ii   What does your answer mean?

4 N people have a meal in a restaurant and decide to share the cost equally.

    The amount they each pay is given by the formula $1.1 \times \dfrac{F + D}{N}$

    where F is the cost of the food and D is the cost of the drink.

    a   Find how much each person pays if:
       i   5 people visit a pizza parlour where the food costs £20 and the drinks cost £5
       ii   A couple visit a posh restaurant where the food costs £44 and the drink costs £24
    b   What does the 1.1 represent?

5   Show that $\dfrac{a + b}{c} = \dfrac{a}{c} + \dfrac{b}{c}$ is an identity by substituting into both sides:

    a $a = 9$, $b = 5$ and $c = 2$ **b** $a = 9x$, $b = 12x$ and $c = 3$

6   Show that $(a - b)^2 \equiv a^2 - 2ab + b^2$ is an identity by substituting into both sides:

    a $a = 9$ and $b = 5$ **b** $a = 5x$ and $b = 2x$

3B Solving problems using equations

Answer each question by constructing and solving an equation.

1 Marcia has 13 more football cards than Derek. They have 89 cards altogether. How many does each person have? Let the number of cards that Derek has be x.

2 Tamsin filled a bowl to the brim using five cartons of juice and 12 cl of lemonade. Justin filled an identical bowl to the brim using three cartons of juice and 60 cl of lemonade. How much juice does a carton contain? Let a carton of juice contain x cl.

3 Fahmida thought of a number, subtracted two, multiplied the answer by three then subtracted the number she first thought of. Her answer was eight. What was the number she first thought of? Let the number be x.

4 Michelle is three times my age. The sum of our ages is the same as six times my age two years ago. How old am I? Let my age now be x.

5 Amy, John and Habib have 210 marbles between them. John has 10 more than Amy. Habib has twice as many as John. How many marbles do they have each? Let the number of marbles that Amy has be x.

3C Equations involving fractions

Solve these equations. Show your working.

1 **a** $\frac{x}{4} = 9$ **b** $\frac{w}{7} = -2$ **c** $\frac{m}{0.4} = 0.3$ **d** $\frac{3d}{5} = 6$

e $\frac{4u}{3} = 3$ **f** $\frac{2p}{7} = 4$ **g** $-\frac{3m}{5} = 9$ **h** $\frac{2t}{-3} = -5$

2 **a** $\frac{s}{8} = \frac{3}{4}$ **b** $\frac{m}{5} = \frac{1}{2}$ **c** $\frac{h}{3} = \frac{4}{5}$ **d** $\frac{3m}{10} = \frac{12}{5}$

e $\frac{2j}{3} = \frac{6}{7}$ **f** $\frac{3t}{2} = \frac{5}{6}$ **g** $\frac{4y}{5} = \frac{1}{20}$ **h** $\frac{2n}{7} = -\frac{3}{5}$

3 **a** $h + \frac{3}{4} = \frac{5}{6}$ **b** $\frac{3}{4}p - 2 = 7$ **c** $\frac{2}{3}a + \frac{1}{3} = 1$

d $\frac{4}{5}g - \frac{3}{5} = \frac{3}{5}$ **e** $\frac{1}{2}x + \frac{3}{8} = 5$ **f** $\frac{3}{10}c - \frac{2}{5} = \frac{1}{2}$

4 **a** $\frac{1}{4}b + \frac{1}{2}b = 6$ **b** $\frac{4}{5}d - \frac{1}{5}d = \frac{7}{10}$ **c** $\frac{1}{3}f + \frac{5}{6}f = 14$

5 **a** $\frac{x+2}{3} = 2$ **b** $\frac{y-6}{5} = 1$ **c** $\frac{2m+3}{3} = 5$

6 **a** $\frac{r+2}{3} = \frac{r+3}{2}$ **b** $\frac{s-3}{2} = \frac{2s+1}{5}$ **c** $\frac{4t-3}{6} = \frac{t-1}{4}$

7 **a** $\frac{1}{h} = 4$ **b** $\frac{10}{y} = 2$ **c** $\frac{2}{w} = 7$ **d** $\frac{4}{p} = -3$

     e $\frac{1}{2e} = 1$ **f** $\frac{5}{3i} = 10$ **g** $\frac{8}{3k} = \frac{5}{6}$ **h** $\frac{4}{5x} = \frac{8}{15}$

Practice

3D Equations involving x^2

For Questions 1–6, solve the equations. There are two solutions for each equation. Show your working.

1 **a** $d^2 = 64$ **b** $x^2 = 169$ **c** $m^2 = 841$ **d** $h^2 = 0.16$

2 **a** $s^2 + 3 = 52$ **b** $g^2 - 20 = 16$ **c** $10 + a^2 = 35$ **d** $101 - y^2 = 20$

3 **a** $5t^2 = 20$ **b** $4n^2 = 196$ **c** $-w^2 = -121$ **d** $1.5f^2 = 54$

4 **a** $\frac{m^2}{2} = 50$ **b** $\frac{u^2}{10} = 40$ **c** $\frac{4p^2}{3} = 12$ **d** $y^2 = \frac{25}{81}$

5 **a** $(x+3)^2 = 100$ **b** $(v-5)^2 = 81$ **c** $(b+10)^2 = 64$ **d** $(2d+3)^2 = 16$

6 **a** $5 = \frac{125}{x^2}$ **b** $3 = \frac{867}{y^2}$ **c** $4 = \frac{1}{u^2}$

Answer Questions 7 and 8 by constructing an equation and solving it.

7 Sarah thought of a number, squared it and added 19. Her answer was 215. What was the number she first thought of?

8 Sean squared a number then tripled it to give 108. What was the number he started with?

Practice

3E Trial and improvement

1 Solve the equation $x^2 + x = 80$, giving your answer correct to 1 decimal place. Copy and complete this table.

x	$x^2 + x$	Comment
8	$8^2 + 8 = 72$	too small
9		
8.5		
8.4		
8.45		
8.46		

2 Solve these equations, giving your answer correct to 1 decimal place. Make a table for each equation.

a $x^2 + x = 40$ **b** $x^2 - x = 66$ **c** $x^3 + x = 7$ **d** $x^3 - x = 20$

Practice

3F Graphs showing direct proportion

1 These tables show the amount charged for different weights of cheddar cheese in two shops.

Country Fare

Weight, W g	200	350	500	600	900
Cost, C pence	84	147	210	243	369

Farm Fresh

Weight, W g	150	400	550	700	1200
Cost, C pence	66	176	242	308	528

a Calculate the ratio $\dfrac{Cost}{Weight}$ for each pair of values.
b For which shop is the cost directly proportional to the weight of cheese? Give a reason for your answer.
c Write an equation connecting C and W for this shop.
d **i** How much would this shop charge for 1.6 kg of cheese?
   ii What weight of cheese can be bought for £5?

2 The diagram shows how two businesses charge for the hire of a canoe.
a For which business is the charge directly proportional to the hire period? Give a reason for your answer.
b For this business, what is the hire charge for 1 hour?

c Write an equation connecting the hire charge, H, with the hire period, p.

d i What would this company charge to hire the canoe for 4.5 hours?

ii What is the hire period corresponding to a hire charge of £9.50? Give your answer in hours and minutes.

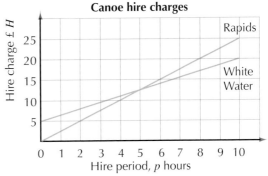

Canoe hire charges

3 Chantal measured the distance a wind-up car travelled after different numbers of winds. This table shows her results.

Number of winds, n	5	10	15	20	30	50
Distance travelled, d metres	3.1	5.7	9.3	11.5	18.7	28.7

a Calculate the ratio $\frac{d}{n}$ for each pair of values.

b Is the distance travelled roughly directly proportional to the number of winds? Give a reason for your answer.

c Write an equation connecting d and n.

d i Estimate how far the car would travel on 25 winds.

ii Estimate the number of winds needed to make the car travel 24 m.

4 The number of bricks, n, to build d metres of wall is given by the ratio:

$n : d = 875 : 25$

a Simplify the ratio $n : d$.

b Write an equation connecting n and d.

c Copy and complete the table.

Length of wall, d metres	5	10	15	20	25
Number of bricks, n					875

d Calculate the number of bricks needed to build 18 m of the wall.

e Draw a graph connecting d and n.

5 The number of badges, N, Yvonne can sew on to T-shirts is directly proportional to the time, t minutes, she spends sewing. It takes her 40 minutes to sew 12 badges.

a Copy, complete and simplify the ratios:

i $\dfrac{\textit{Number of badges}}{\textit{Time taken}} =$ ii $N : t =$

b Write an equation connecting N and t.

c i How many badges can Yvonne sew in 70 minutes?

ii How long does it take Yvonne to sew 39 badges?

CHAPTER 4 Geometry and Measures 1

Practice

4A Angles of polygons

1 A polygon has 20 sides.

   a How many triangles does the polygon contain?
   b What is the sum of its interior angles?

2 The sum of the interior angles of a polygon is 2520°.

   a How many triangles does the polygon contain?
   b How many sides does the polygon have?

3 The interior angles of a hexagon are 100°, 150°, 75°, 90°, 145° and x.
   Find x.

4 Calculate the size of the unknown interior angle.

a

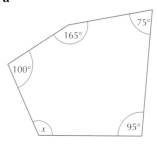

b

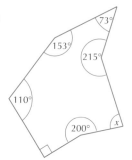

c

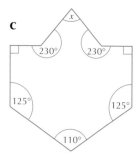

5 Calculate the size of the unknown exterior angle.

a

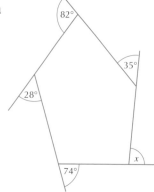

b

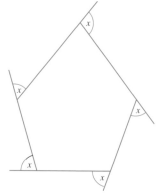

6 Find interior angle *x*.

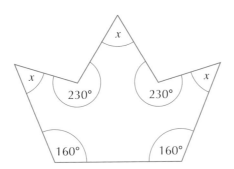

7 **a** Calculate the marked angles. Explain how you found each angle.

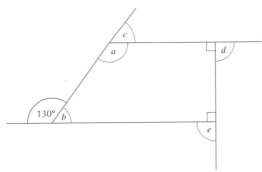

b Check that the exterior angles total 360°.

Practice **4B Angles of regular polygons**

1 Find the sum of the exterior angles of a regular polygon with 13 sides.

2 Copy and complete this table for regular polygons.
Give reasons for your answers.

Number of sides	Sum of exterior angles	Size of each exterior angle	Size of each interior angle
15			
16			
18			
20			
30			

3 **a** Calculate the marked angles for this regular hexagon.
 b Prove that ABDE is a rectangle.
   Hint: Show that its angles are all 90°.

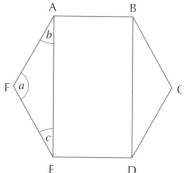

4 Calculate the number of sides for a polygon whose:

   a exterior angle is 4°.
   b interior angle is 171°.
   c interior angles add up to 1980°

5 Calculate the marked angles for this regular octagon.

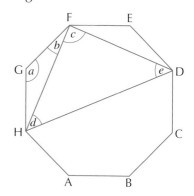

6 This diagram shows a regular pentagon and regular hexagon
   joined together.
   Calculate the marked angle.

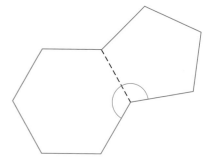

5

1 a Draw a circle with radius 33 mm.
   b Draw a circle with diameter 9.2 cm.
   c Draw a semicircle with diameter 8 cm.
   d Draw a quadrant of a circle with radius 52 mm.

2 Construct these diagrams.
   Hint: To find out the radius of each circle, measure the diameter, and halve it.

a

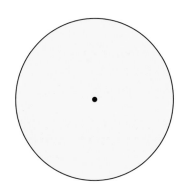

b

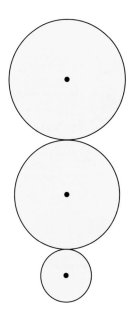

3 Construct these diagrams.

a

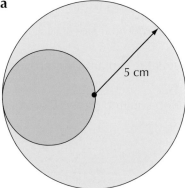

5 cm

b

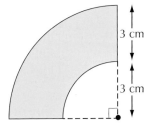

3 cm

3 cm

c

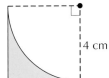

4 cm

4 **a** Draw a regular octagon.

Hint: Draw the circle first, then the diameters, then complete the polygon.

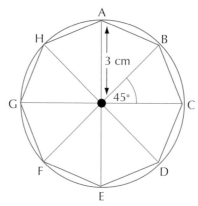

b Measure the length of the side of the polygon.

Practice

4D Tessellations and regular polygons

1 **a** Copy each triangle on to squared paper.
Does each triangle tessellate?

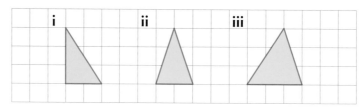

b Draw a scalene triangle (three unequal sides) of your own on squared paper. Does it tessellate?
c Does a triangle always tessellate? Explain your answer.

2 **a** Copy each quadrilateral on to squared paper. Does each quadrilateral tessellate?

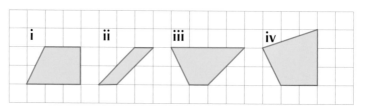

b Draw a quadrilateral of your own on squared paper (not a special quadrilateral). Does it tessellate?
c Does a quadrilateral always tessellate? Explain your answer.

4E Constructing triangles

Show all of your construction lines.

1. Construct these right-angled triangles.

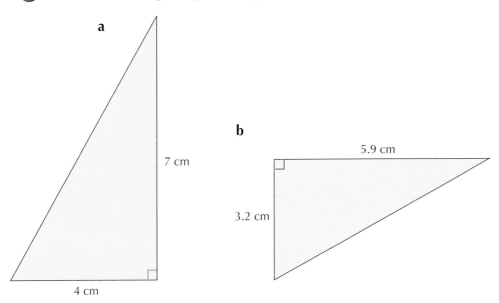

a

7 cm

4 cm

b

5.9 cm

3.2 cm

2. Construct a right-angled triangle with base 8 cm and height 7 cm.

3. **a** Construct an accurate scale drawing of this playground slide.
   Use a scale of 1 cm to 1 m.

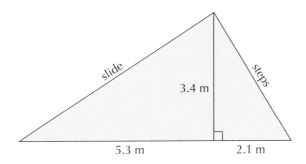

slide

3.4 m

steps

5.3 m

2.1 m

   b Find the real length of the slide.

4 Construct these triangles.

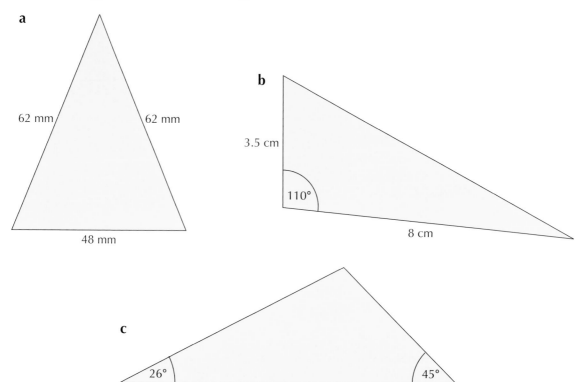

Practice

4F Loci

1 Draw a line AB that is 9 cm long. Use a ruler and compasses only to construct the perpendicular bisector of AB. Mark C as the midpoint of AB.

2 Use a ruler and compasses only to construct the angle bisector of this angle.

FM **3** **a** A radio controlled airship, A, has a range of 400 m from the radio control unit, X. Construct a scale drawing of the locus of the airship using a scale of 1 cm to 100 m.

b A second airship, B, has a range of less than 300 m from its radio control unit Y. The radio controllers X and Y are 500 m apart. Shade the region where the two airships could collide.

c Airship, A, steers a course equidistant from the two radio controllers. Draw its locus.

FM ④ This diagram shows a cross-section of a valley APB. There is a helicopter landing pad at P.

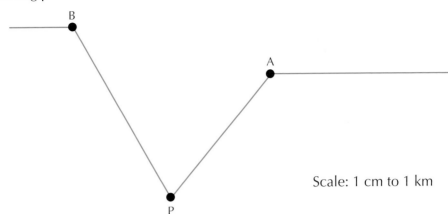

Scale: 1 cm to 1 km

a Trace the diagram.

b A helicopter, H, flies toward the landing pad, keeping equidistant from beacons A and B. Construct its locus using a dotted line.

c A straight laser tracking beam is continuously transmitted from the landing pad, equidistant from the walls of the valley. Construct its locus using a dotted line.

d When the helicopter reaches the beam, it changes course and flies straight to the landing pad. Use a solid line to draw the flight path of the helicopter.

e Mark the two points, X, where the helicopter is 4 km from beacon A.

Practice
4G Geometrical reasoning

1 Describe the convention used in diagrams to show each of these features. Use diagrams to illustrate your answers, where necessary. There may be more than one convention.

   a Equal sides of a shape
   b Parallel lines
   c The vertices of a triangle
   d A line of symmetry of a shape
   e The length of the base of a triangle
   f The area of a shape
   g The horizontal axis for a straight-line graph
   h The angles of the triangle ABC
   i The angles of an isosceles triangle
   j The image of the shape P when it is reflected in the x-axis

2 Why is this statement *not* a definition of a cylinder?
   'A cylinder is a solid with a circular cross-section.'

3 a Give a definition of a kite.
   b Give a definition of an arrow.

4 A parallelogram is a quadrilateral with equal opposite sides.
   Write down two other definitions of a parallelogram.

5 A regular hexagon is a shape with six equal sides and six equal angles. Write down three derived properties. Use diagrams to illustrate your answers.

6 A triangle that has exactly two equal sides is called isosceles. Write down two derived properties.

7 Describe each of these as a convention, definition or derived property.

    a The set of points equidistant from a fixed point is called a circle.
    b The diameter of a circle is twice its radius.
    c The centre of a circle is labelled O.

8 This diagram shows triangle ABC labelled clockwise and anti-clockwise. Is there a convention? Investigate by looking in your pupil book and other textbooks.

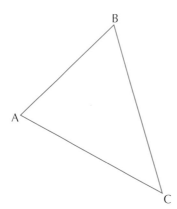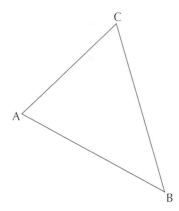

CHAPTER **5** Statistics **1**

Practice

5A Scatter graphs and correlation

1 There are three ways of combining the variables x, y, z, in pairs (xy, yz and zx). Write down what sort of correlation each pair will have.

    x: The number of people who travel to work by car each month
    y: The number of people who travel to work by public transport each month
    z: The amount of car exhaust pollution each month

Illustrate each correlation by sketching a scatter graph.

2 Ten identical houses had their lofts converted. These scatter graphs show the number of workers, time taken and cost of equipment hire for each loft.

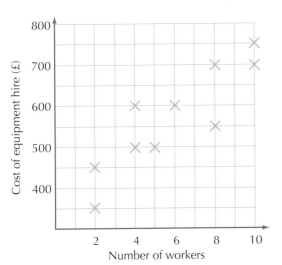

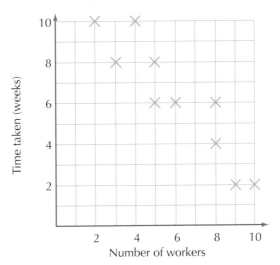

a   Describe the correlation between the number of workers and time taken.
b   Describe the correlation between the number of workers and the cost of equipment hire.
c   i   Describe the correlation between the time taken and the cost of equipment hire.
   ii   Sketch a scatter graph to illustrate this correlation.

3 This table shows the size and price of books sold by a bookshop in an hour.

Number of pages, n	60	60	120	140	140	180	240	280	320	400	480
Price of book, £P	1.50	1.00	2.50	3.00	3.20	5.00	4.50	6.00	6.00	7.50	10.00

a   Draw a scatter graph for the data. Use these scales.
   Number of pages: 2 cm to 100 pages
   Price of book: 1 cm to £1
b   Draw a line of best fit, by eye.
c   Use your line of best fit to estimate:
   i   the cost of a book containing 260 pages.
   ii   the number of pages in a book costing £1.50.

This diagram shows how the US produced its energy during the last century.

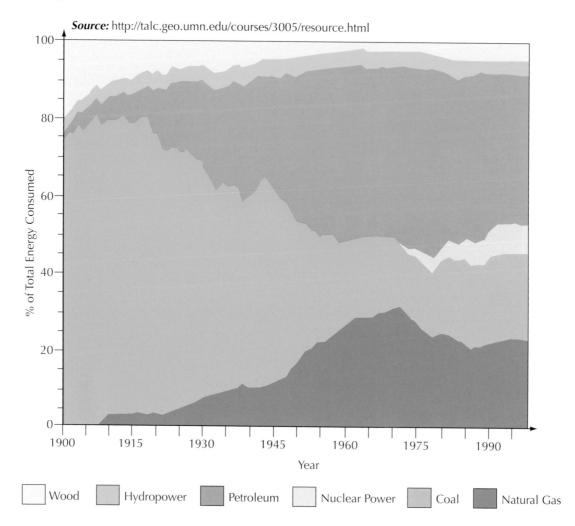

Source: http://talc.geo.umn.edu/courses/3005/resource.html

Legend: Wood | Hydropower | Petroleum | Nuclear Power | Coal | Natural Gas

The height of each stripe is the percentage of the total energy that the power source provides.
For example, in 1900, hydropower provided about 4% of the total energy in the US.

a What was the main source of energy at the:
  i beginning of the century?
  ii end of the century?

b Which source of energy provided roughly the same percentage of energy throughout the century?

c Estimate when petroleum and coal were used in equal amounts.

d What percentage of US energy was produced from natural gas in 1975?

e What percentage of US energy was produced from coal in 1945?

f Describe the trend of using wood as an energy source.

g Which energy source provided about twice the energy as hydropower at the end of the century?

h Describe the trend in the use of natural gas.

i The total energy consumed in 1990 was 82 quadrillion units.
  a Find out what a quadrillion is.
  b Estimate the energy produced from petroleum in 1990.

2 These graphs show the prices and sales of shares in two companies, James, James and Yates (JJY) and Standard Wire (SW), during a day of trading. The number of shares sold is called the 'volume'.

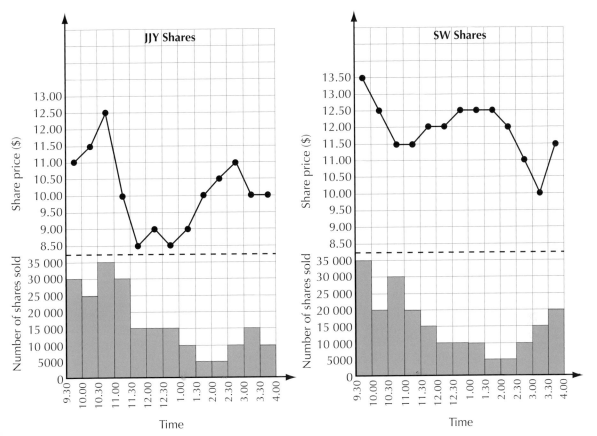

The prices shown are at the end of each 30-minute period. For example, the price of SW was $11.50 at 4 pm, and 20 000 shares were sold between 3.30 pm and 4 pm.

a Which share started trading at the highest price of the day?

b Which share has the greatest range of prices?

c What is the greatest price increase during a half-hour period?
State the share and when it occurred.

d Bad news caused one of the shares to greatly reduce in price.
  i Which share was this?
  ii When did this happen?
  iii What was the fall in price?

e What was the difference in share prices at 12.30 pm?

f What do you notice about the volume during the day for both shares?
Give a reason why this might have happened.

g During which hour was the least number of shares traded for each company?

h How much money was spent on JJY shares from 2.30 pm to 3.00 pm?

i During which half-hour period was most money spent altogether?
How much was spent?

5C Two-way tables

 1 This table show some holiday prices (£) to Tenerife and Gran Canaria.

	Tenerife						Gran Canaria					
	1 week			2 weeks			1 week			2 weeks		
Date	S/C	H/B	A/I	S/C	H/B	A/I	S/C	H/B	A/I	S/C	H/B	A/I
28 Feb	159	269	335	200	320	375	177	309	319	225	349	350
10 Mar	159	275	339	200	325	377	189	309	319	220	349	359
14 Mar	170	285	349	208	329	375	190	315	335	218	340	339
21 Mar	175	289	349	215	340	380	190	319	335	220	340	339
28 Mar	149	279	339	215	320	380	195	330	339	216	365	345
4 Apr	169	269	345	219	315	375	199	389	379	225	419	399

S/C = self-catering; H/B = hotel, half-board; A/I = all-inclusive

a What is the most expensive two-week holiday in March?

b What is the range of prices of one-week holidays?

c Overall, from which date is it most expensive to travel to Tenerife?

d Describe any patterns you notice.

e Make a two-way table showing the difference in price between one-week and two-week holidays to Gran Canaria. What do you notice?

f Which type of holiday to Tenerife (S/C, H/B or A/I) varies in price the least? Explain your answer.

 2 This table shows the percentage of the population owning a computer in 1999 and 2000, and also the percentage of the population owning a computer connected to the Internet.

		1999		2000	
		Owns a computer connected to the Internet (%)	Owns a computer (%)	Owns a computer connected to the Internet (%)	Owns a computer (%)
Age	15–24	13	45	48	62
	25–34	14	47	51	61
	35–44	17	52	62	71
	45–54	15	48	53	61
	55–64	10	30	35	43
	65–74	4	15	13	20
	75+	2	7	7	10

Source for 2000 data: www.oftel.gov.uk/publications/research
Data for 1999 was estimated based on average percentage change from 2000.

a Describe any trends you notice.

b Compare 1999 with 2000.

c Construct a two-way table showing the percentages of people owning a computer not connected to the Internet.

6

5

6

Practice 5D Statistical investigations

Conduct an investigation into one of these areas. Follow the steps given in the table in statistical investigation section of chapter 5 of Pupil Book 2.

FM **1** Typing

FM **2** Computers at home

FM **3** Boiling an egg

CHAPTER **6**
Geometry and Measures **2**

Practice 6A Circumference of a circle

1 **a** Name the lines on this diagram.
  i AB   **ii** AC
  iii DE   **iv** CF
  v CG
 b Name these shapes.
  i ABCA   **ii** ACFA
  iii ABCFA **iv** GABCG

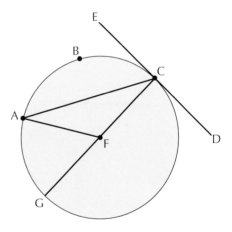

2 **a** Measure the diameter of a bicycle wheel (or another circle, such as a saucepan lid) to the nearest centimetre.
 b Measure the circumference of the wheel by rolling it along the ground for one turn.
 c Use a calculator to divide your answer to **b** by your answer to **a**.
 d The answer should be a little more than 3. How much more?

Practice

6B The formula for the circumference of a circle

1 Use the approximation $\pi \approx 3.14$ to calculate the circumference of each wheel. Write your answers correct to the nearest centimetre.

a

60 cm

b

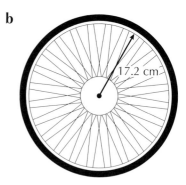

17.2 cm

2 Use the π button of your calculator to calculate the circumference of each coaster. Write your answers correct to 1 decimal place.

a

10 cm

b

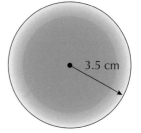

3.5 cm

3 Use the approximation $\pi \approx \frac{22}{7}$ to calculate the circumference of each coin. Do not use a calculator.

a

TWO POUNDS

14 mm

1998

b

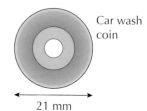

Car wash coin

21 mm

4 Calculate the total length of the lines in this crop circle. It has two semicircles, a circle and straight lines. Write your answers correct to the nearest centimetre.

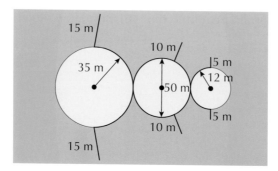

15 m

10 m

35 m

5 m

12 m

50 m

10 m

5 m

15 m

6

Use the approximation π ≈ 3.14 or the π button of your calculator for
these questions.

1 Calculate the area of the lid of each tin. Write your answers using a suitable
degree of accuracy.

a

b

c

2 Calculate the total area of this
arched window.
Write your answer correct to the
nearest square centimetre.

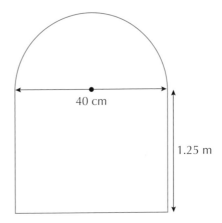

3 This diagram shows how a washer is made.

  a Calculate the area of the blank, correct to 1 decimal place.
  b Calculate the area of the hole, correct to 1 decimal place.
  c Calculate the area of the finished washer, correct to the nearest mm².

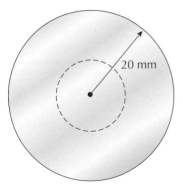

Blank

Centre removed

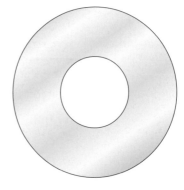

Finished washer

6D Metric units for area and volume

5

1. Convert these units.

   a 9 m² to cm²
   b 0.05 m² to cm²
   c 5000 cm² to m²
   d 17 cm² to m²
   e 11 cm² to mm²
   f 850 mm² to cm²
   g 0.065 m² to mm²
   h 28 400 mm² to m²
   i 75 ha to m²
   j 0.085 ha to m²
   k 39 000 m² to ha
   l 1 ha to mm²

2. Convert these units.

   a 9 cm³ to mm³
   b 0.066 cm³ to mm³
   c 3 m³ to cm³
   d 0.0004 m³ to cm³
   e 650 mm³ to cm³
   f 22 500 cm³ to m³
   g 6 m³ to mm³
   h 1 400 000 mm³ to m³

3. Convert these units.

   a 72 litres to ml
   b 0.15 litres to ml
   c 700 ml to litres
   d 27 500 ml to litres
   e 4.2 litres to cm³
   f 4000 cm³ to litres
   g 39 cm³ to ml
   h 5.2 m³ to litres
   i 80 cm³ to litres
   j 4300 ml to cm³
   k 24 000 litres to m³
   l 850 litres to m³

6

4. A rectangular aircraft hangar measures 940 m by 750 m. Find its area, giving your answer in hectares.

5. A rectangular stamp measures 32 mm by 34 mm.

   a What is the area of the stamp in **i** mm² **ii** cm²?
   b A sheet of stamps has an area of 1.632 m². How many stamps are there?

FM 6. An inflatable dinghy has a capacity of 0.66 m³. Martha's pump blows 1.2 litres of air into the dinghy every second. How long will it take her to fill the dinghy?

Practice

6E Volume and surface area of prisms

5

1. Convert these quantities.

   a 0.07 m² to cm²
   b 8400 cm³ to l
   c 21 000 000 cm³ to m³
   d 4.8 ml to mm³
   e 5340 mm² to cm²
   f 5200 l to m³

34

2 **i** Calculate the surface area of each prism.
    ii Calculate the volume of each prism.

a

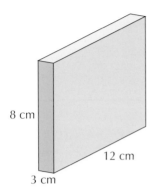

8 cm

12 cm

3 cm

b

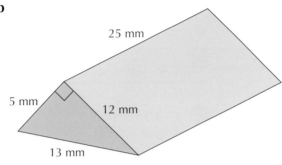

25 mm

5 mm

12 mm

13 mm

3 Convert your answers to Question **2b** to cm² and cm³.

4 The cross-section of this plastic bench has an area of 620 cm². Calculate its volume in **a** cm³ and **b** m³.

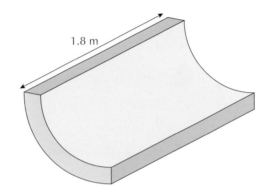

1.8 m

5 The diagram shows a petrol tank in the shape of a prism.

a Calculate the area of the cross-section.
b Calculate the volume of the tank.
c Calculate the capacity of the tank in litres.
d Calculate the surface area of the tank in m².

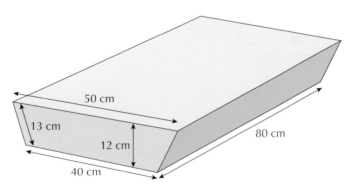

50 cm

13 cm

12 cm

80 cm

40 cm

CHAPTER 7 Number 2

Practice

7A Powers of 10

1   **a** 4.15×100 **b** $3.91 \div 100$ **c** 34×1000
   d $5.9 \div 1000$ **e** 2.3×0.1 **f** $0.52 \div 0.1$

2   Multiply these numbers by **i** 10^2 and **ii** 10^4.

   a   0.42   **b**   874   **c**   12.6
   d   0.053   **e**   0.0004

3   Divide these numbers by **i** 10^2 and **ii** 10^3.

   a   4300   **b**   0.6   **c**   23.7
   d   0.054   **e**   13 599

4   Multiply these numbers by **i** 0.01 and **ii** 0.0001.

   a   600   **b**   5   **c**   23 000
   d   0.6   **e**   25.2

5   Divide these numbers by **i** 0.1 and **ii** 0.001.

   a   7   **b**   0.54   **c**   98
   d   0.012   **e**   492   **f**   0.0087

6   Calculate these.

   a 5.25×10^3 **b** $0.43 \div 0.01$ **c** $7450 \div 10^2$
   d 25×0.01 **e** 0.003×10^4 **f** $63 \div 0.001$

7   Calculate these.

   a 7000×10^{-2} **b** 4.5×10^{-1} **c** $72 \div 10^{-1}$
   d $0.4 \div 10^{-2}$ **e** 250×10^{-3} **f** $9 \div 10^{-3}$

8   Copy and complete each ladder of calculations. The first one has been started for you.

a

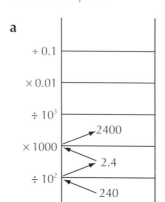

b

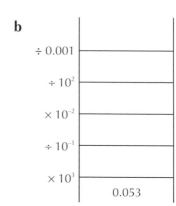

Practice — 7B Rounding

1 Round these numbers **i** to 1 decimal place and **ii** to 2 decimal places.

   a 2.874   **b** 0.05537   **c** 20.957

   d 0.5059   **e** 9.98407

2 Round these numbers to 1 significant figure.

   a 327   **b** 8.452   **c** 8175   **d** 0.06594

   e 79 550   **f** 0.007 29   **g** 20.981

3 Estimate answers to these by first rounding the numbers to 1 significant figure.

   a 0.76×47 **b** 12% of \$2761

   c $38\ 297 \div 77$ **d** $5.38(13.85 - 4.199)$

   e $35.44^2 - 22.2^2$ **f** $0.9167 \div 0.0286$

   g $\dfrac{5.018 \times 3.86}{11.41 - 6.459}$ **h** $\dfrac{37.71 + 18.099}{0.829 - 0.333}$

4 Round these quantities to an appropriate degree of accuracy.

   a Jeremy is 1.8256 m tall.

   b A computer disk holds 688 332 800 bytes of information.

   c A petrol tank has a capacity of 72.892 litres.

   d An English dictionary contains 59 238 722 words.

   e Kailash held his breath for 82.71 seconds.

5 Use a calculator to work out these.
Round your answers to a suitable degree of accuracy.

   a 7392^2 **b** $\dfrac{5}{7}$

   c $18.3 + 2.8 \times 9.7 \times 1.4$ **d** $\dfrac{997 - 285}{0.863 \times 5.25}$

Practice — 7C Multiplying decimals

Do not use a calculator. Show your working.

1 **a** 0.4×0.2 **b** 0.04×0.2 **c** 0.7×0.6 **d** 0.7×0.06

    e 0.09×0.9 **f** 0.9×0.9 **g** 0.04×0.3 **h** 0.8×0.03

2 **a** 300×0.7 **b** 0.2×70 **c** 80×0.5 **d** 0.7×800

    e 400×0.01 **f** 0.06×500 **g** 2000×0.03 **h** 0.002×6000

3 **a** 7×2.6 **b** 0.8×7.1 **c** 5.51×0.2 **d** 0.3×9.25

4 Rice costs £2.15 per kg. Calculate the cost of 0.8 kg of rice. Work in £.

5 The tank of a model aeroplane holds 8.28 cl of fuel. How much fuel is needed to fill the tank 6 times?

7D Dividing decimals

Do not use a calculator. Show your working.

1　**a**　$0.84 \div 0.2$　　**b**　$0.49 \div 0.7$　　**c**　$0.24 \div 0.8$　　**d**　$0.66 \div 0.1$

2　**a**　$50 \div 0.2$　　**b**　$200 \div 0.8$　　**c**　$30 \div 0.6$　　**d**　$900 \div 0.3$

3　**a**　$5.2 \div 4$　　**b**　$2.8 \div 70$　　**c**　$72 \div 80$　　**d**　$9.6 \div 60$

4　**a**　$6.4 \div 32$　　**b**　$5.5 \div 2.5$　　**c**　$2.7 \div 1.8$　　**d**　$25.5 \div 3.4$

5　30 buttons cost 228p. Calculate the cost of one button.

6　How many stamps fit across the top of the envelope?

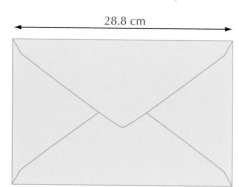

28.8 cm

1.8 cm

7E Efficient use of a calculator

Round your answers to a suitable degree of accuracy, where necessary.

1　Use the power key on your calculator to evaluate these numbers.

　a　6^6　　　　**b**　0.81^3　　　　**c**　1.27^5　　　　**d**　230^3

2　Use the fraction key to calculate these.

　a　$\frac{7}{10} + \frac{9}{20} - \frac{3}{5}$　　**b**　$2\frac{1}{3} \div 1\frac{8}{9}$　　**c**　$\frac{3}{5} \times \left(\frac{7}{9} - \frac{2}{3}\right)$　　**d**　$\left(3\frac{2}{3}\right)^2$

7 CH

7

3 Calculate these.

a $2.5^2 \times (1.8 + 5.4)$ **b** $3^3 - 2.5^2$ **c** $(12^3 + 17 + 25) \div 30$

d $\sqrt{0.9^2 + 0.6^2}$ **e** $\sqrt{13 - \frac{7}{11}}$

4 Calculate the perimeter of the dinner mat.

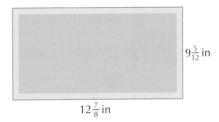

$9\frac{5}{12}$ in

$12\frac{7}{8}$ in

5 **a** Find the symbol $x!$ on your calculator. Use this to work out:

i $1!$

ii $2!$

iii $3!$

b Copy and complete this table.

x	1	2	3	4	5
$x!$					

c Describe any pattern you notice.

d Use your pattern to find the answer to $6!$.
Check your answer using a calculator.

e Copy and complete these calculations.
$1 \quad\quad\quad = 1$
$2 \times 1 \quad\quad = 2$
$3 \times 2 \times 1 \quad =$
$4 \times 3 \times 2 \times 1 =$

f What do you notice about the answers?

g Use your calculator to work out:

i $9 \times 8 \times 7 \times 6 \times 5 \times 4 \times 3 \times 2 \times 1$
ii $12 \times 11 \times 10 \times 9 \times 8 \times 7 \times 6 \times 5 \times 4 \times 3 \times 2 \times 1$

7

 FM   ① Jonathan has been on a diet. He wants to weigh less than 70 kg.
He used some old weighing scales and found his weight was 155 lb.
Does he need to stay on the diet? Use the conversion: 1 kg = 2.2 lb.

 FM   ② Which biscuit purchase gives the best value? Explain your answer.

ⓐ

Crumblies   ⟩74p⟨   20 biscuits
BISCUITS

ⓑ

Crumblies *extra value* ⟩£1.19⟨ 35 biscuits
BISCUITS

ⓒ

Crumblies
BISCUITS
⟩£1.26⟨   *Family Pack*   36 biscuits

③ The total capacity of 5 cups and 3 mugs is 230 cl. The total capacity of 5 cups and 4 mugs is 265 cl.

  a Find the capacity of a mug.
  b Find the capacity of a cup.

④ Arooma thinks of a number, adds 4 then doubles the result. Her answer is 42. What was the number she first thought of?

⑤ The diagram shows the heights of two trees. During the next year, the bamboo tree grew 10% and the willow tree grew 5%. Which is the taller tree now?

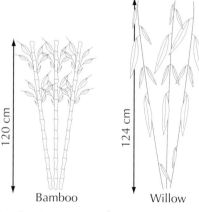

⑥ The TV programme Wacko Magic lasts 25 minutes and is shown every Tuesday and Thursday.
Country Facts lasts 15 minutes and is shown every Wednesday.
Karen records these programmes each week. She has recorded 7 hours and 35 minutes on a videotape.

  a How many weeks did she record?
  b How many of each programme did she record?

CHAPTER 8 Algebra 4

8A Factorisation

1 i Write down the common factors of each pair of numbers.
ii Find the HCF of each pair of numbers.

    a 8 and 20   **b** 9 and 12   **c** 14 and 28 **d** 12 and 18

2 Find the HCF for each set of numbers.

    a 30 and 45   **b** 22 and 66   **c** 48 and 64
    d 8, 12, 16   **e** 10, 20, 35   **f** 6, 12, 24

3 Write down all of the factors of each expression.

    a $3n$ **b** de **c** $4w$ **d** $7st$
    e $10mn$ **f** g^2 **g** $6k^2$ **h** x^2y

4 i Write down the common factors for each pair of expressions.
ii Find the HCF.

    a 8 and $4c$ **b** $6d$ and $4e$ **c** fg and fh
    d $9p$ and $6p$ **e** $2ab$ and $5ac$ **f** $10ew$ and $15eb$
    g $8u$ and $12ut$

5 Find the HCF of each pair of expressions.

    a $9mn$ and $10m$ **b** de and ce **c** $6y$ and $9t$
    d $7pq$ and $14pr$ **e** $12x$ and $16wx$ **f** $20a$ and $30ab$

6 Find the HCF of each pair of expressions.

    a m^2 and m **b** $6n^2$ and $5n$ **c** x^2y and xy
    d $9d^2$ and $6d$ **e** $10m^2n$ and $15m$ **f** s^2t^2 and st^2
    g $16g^2h$ and $20gh^2$

7 Find the HCF of each set of expressions.

    a $4a$, 10, 12   **b** mn, mr, ms **c** $6t$, $12t$, $18t$
    d $10pq$, $5pr$, $20pq$ **e** $8ab$, ab, $4ac$ **f** $14hi$, $7h$, $28hi$
    g $6wx$, $9wy$, $12xy$

8B Index notation with algebra

1 Calculate these powers. Use a calculator where necessary.

    a 2^6 **b** 3^3 **c** 10^7
    d 9^4 **e** 11^5 **f** 8^9

2 Write these using index notation.

    a $d \times d \times d \times d$ **b** $s \times s \times s \times s \times s \times s$

For Questions 3–7, simplify the expressions.

3
  a $5a \times a \times a$ **b** $4y \times 3y \times y \times y \times y$
  c $2g \times g \times 4g \times 2g$ **d** $3z \times 3z \times 3z \times 3z \times 3z$
  e $s \times s \times 3s \times 2s \times s \times 5s$ **f** $(2n)^3$
  g $(4d)^6$

4
  a $a^2 \times a^2$ **b** $m^4 \times m^2$ **c** $u \times u^3$
  d $p^2 \times p^7$ **e** $t^2 \times t^3 \times t$

5
  a $2y^2 \times 3y$ **b** $5a^3 \times 2a^2$ **c** $e^2 \times 6e^3$
  d $9c \times 3c^2 \times c^3$ **e** $m^2 \times 2m^4 \times 5m^3$

6
  a $\dfrac{e^5}{e}$ **b** $\dfrac{k^4}{k^2}$ **c** $p^7 \div p^4$ **d** $\dfrac{h^5}{h^4}$ **e** $\dfrac{m^{10}}{m^6}$

7
  a $\dfrac{5w^3}{w}$ **b** $\dfrac{4b^5}{b^2}$ **c** $\dfrac{12c^4}{4}$ **d** $\dfrac{9d^5}{3d^2}$ **e** $\dfrac{20x^7}{5x^3}$

Practice

8C Square roots and cube roots

1 Calculate these, correct to 1 decimal place where necessary.
  a $\sqrt{169}$ **b** $\sqrt{10}$ **c** $\sqrt[3]{343}$
  d $\sqrt[3]{900}$ **e** $\sqrt{5240}$ **f** $\sqrt[3]{-8000}$

2 Calculate these, correct to 2 significant figures where necessary.
  a $\sqrt{2\,000\,000}$ **b** $\sqrt[3]{216\,000}$
  c $\sqrt{40\,000}$ **d** $\sqrt[3]{-150\,000}$

3 Darren made some mistakes in his homework. He wrote cube roots ($\sqrt[3]{\ }$) as square roots ($\sqrt{\ }$). Copy and correct his homework.
  a $\sqrt{64} = 4$ **b** $\sqrt{16} = 4$ **c** $\sqrt{-8} = -2$
  d $\sqrt{20} \approx 4.47$ **e** $\sqrt{90} \approx 4.48$

4 Calculate the value using your calculator. Give your answers to 1 decimal place.
  a $\sqrt{300}$ **b** $\sqrt[3]{300}$ **c** $\sqrt{60}$
  d $\sqrt[3]{60}$ **e** $\sqrt{6}$ **f** $\sqrt[3]{6}$

5 Solve these quadratic equations, correct to 1 decimal place.
  a $x^2 = 40$ **b** $y^2 = 200$ **c** $4h^2 = 20$ **d** $\dfrac{d^2}{7} = 6$

6
  a **i** Calculate the square root of 100.
    ii Calculate the cube root of the answer.
  b **i** Calculate the cube root of 100.
    ii Calculate the square root of the answer.
  c What do you notice? Write down a rule.
  d Test your rule using a different number.

Practice **8D Graphs involving time**

  1 **i** Match each graph with a description.
ii Copy each graph and label the axes.

A
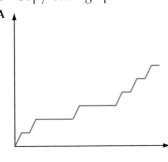

B
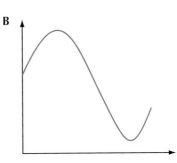

C
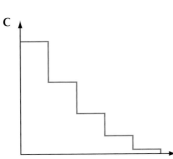

D
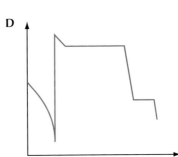

a The petrol in a car tank over time.

b The amount of electricity used by a kettle during a day.

c The height of a swing over time.

d The number of cubes needed to build each level of a pyramid.

  2 This diagram shows the journeys of a submarine travelling at a constant speed.

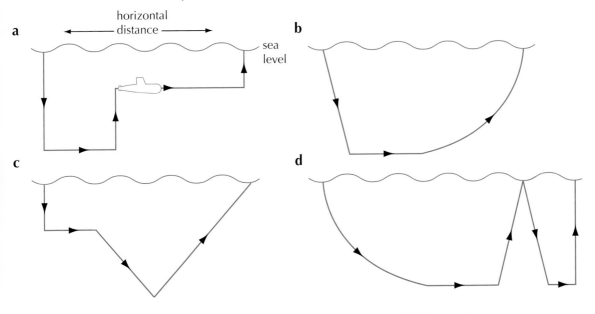

Sketch graphs for the horizontal distance travelled over time.

7

FM ③ An inkjet printer prints letters from left to right and uses ink as shown.

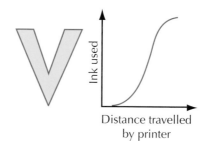

a Sketch graphs to show the relationship between the volume of ink used and the distance travelled by the printer for these letters.

i   ii   iii  iv  v

b Another printer prints from bottom to top. Sketch more graphs for this printer.

FM ④ Sketch a graph to show the height of a cornfield during a year.

CHAPTER 9 Statistics **2**

Practice

9A Probability statements

① Which statements are incorrect? Explain why.

a I never get a seat on the train, so it is unlikely I'll get one today.
b There are two ways a drawing pin can land when it is dropped. So there is an evens chance of it landing point up.

c My name is Andrew. There must be another person in my school named Andrew.
d You could never win the lottery three times in a row.
e Today is 21st August and it is sunny. It will probably be sunny tomorrow.
f Three boxes contain 6, 6, and 12 eggs. One of the eggs is broken. Emma buys one of the boxes. There is a 1 in 3 chance that her box contains the broken egg.
g Sam walked under a ladder. Soona said, 'That's bad luck. Something will probably go wrong for you today.'

2 Which of these pairs of events are *not* independent? Give a reason for your answer.

    a A bird comes into a garden. A second bird joins the first.

    b Choosing a heart from a pack of 52 cards. Choosing a heart from a pack of 52 cards with the four kings removed.

    c I have a box of old pens. I choose a red one and it runs out the same day. I choose a blue one and it runs out the same day.

    d I don't guess your lucky number the first time. I guess your lucky number the second time.

3 These dice have the letters shown on their sides.

Dice 1
6 sided:
A D E E E F

Dice 2
8 sided:
A B B B B D E F

Dice 3
12 sided:
A A B B D D E E F F G G

Dice 4
20 sided:
A A A D D D D E E E E E F F F F F F I I

All the dice are rolled.

    a Which dice has the greatest chance of showing an E?

    b Which dice has the least chance of showing one of the letters D or F?

    c Which dice are unlikely to show a vowel?

    d Which dice is most likely to show one of the letters of the word BAD?

    e Which dice has a fifty-fifty chance of showing a consonant?
       Note: a consonant is a letter that is not a vowel.

    f Which dice has the greatest chance of not showing a letter from the word DEAF?

Practice   **9B Mutually exclusive and exhaustive events**

For Questions 1–3, a ring is randomly thrown at this board and lands on one of the numbered hooks.

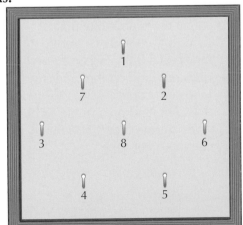

Ring

6

1 Decide whether each pair of events is mutually exclusive or not. Give reasons for your answers.

    a Ring lands on an odd number.
       Ring lands on 8.
    b Ring lands on a prime number.
       Ring lands on a multiple of 3.
    c Ring lands on a square number.
       Ring lands on a prime number.
    d Ring lands on a number greater than 4.
       Ring lands on a factor of 8.

2 **a** Why is this set of events exhaustive?
       Ring lands on the number 1.
       Ring lands on a prime number.
       Ring lands on an even number greater than 3.
    b Write down another set of events that are exhaustive.

3 Calculate the probability that the ring:

    a lands on a prime number.
    b does not land on a square number.
    c lands on a multiple of 3 or an odd number.

4 The dartboard shows the probability of Darren hitting one of five areas. Darren throws a dart.

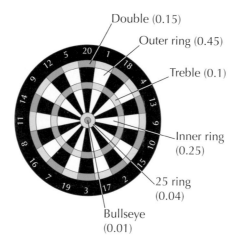

Double (0.15)
Outer ring (0.45)
Treble (0.1)
Inner ring (0.25)
25 ring (0.04)
Bullseye (0.01)

    a Calculate the probability that it:
       i lands on a double or treble.
       ii lands on a bulls eye or 25 ring.
       iii does not land on the inner ring.
       iv lands on a double, outer ring or treble.
       v does not land on a treble or bulls eye.
    b If Darren throws a dart 500 times, how many times would you expect it to hit the outer ring?

Practice

9C Estimates of probability

1 These five identical buttons were dropped 80 times.

The number of buttons that fell white side up was recorded each time.
This table summarises the results.

Number of buttons falling white side up	0	1	2	3	4	5
Frequency	3	14	24	27	11	1
Relative frequency		0.175				

a Copy and complete the table.
b What is the estimated probability of:
  i 3 buttons falling white side up?
  ii no buttons falling white side up?
  iii more than 3 buttons falling white side up?
c If the five buttons were dropped 300 times, how many times would you expect:
  i 3 buttons to fall white side up?
  ii no buttons to fall white side up?
d **i** How many buttons were dropped altogether?
  ii How many buttons landed white side up altogether?
  iii Estimate the probability of a single button landing white side up when dropped.
  iv If you dropped a button to see who goes first in a game, would that be fair?

2 A company that manufactures computer chips makes regular quality control checks. This table shows how many computer chips are checked and how many of them are faulty.

	Hourly check	Daily check	Weekly check	Monthly check
Number of chips checked	20	50	200	500
Number of chips faulty	3	8	26	72
Relative frequency				

a Copy and complete the table.
b Estimate the probability of a computer chip being faulty.
c A batch contains 6000 computer chips. Estimate the number that are faulty.
d Every year the company produces 70 million computer chips. Estimate the number that are faulty.

Geometry and Measures 3

CHAPTER 10

10A Enlargements

7

1 Trace each shape with its centre of enlargement O. Enlarge the shape by the given scale factor.

a

Scale factor −3

b

Scale factor −2

c

Scale factor $\frac{1}{3}$

d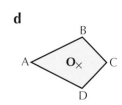

Scale factor $\frac{1}{2}$

2 Copy each diagram and enlarge the shape about the point C using the given scale factor.

a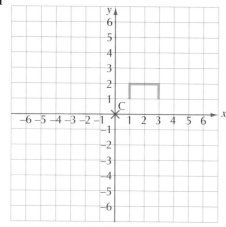

Scale factor −2

b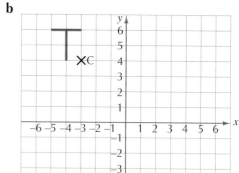

Scale factor −4

3 Shape A′B′C′D′ is the enlargement of shape ABCD using a scale factor of −2 about (0, 0). Find the points A, B, C and D using rays and draw the shape ABCD.

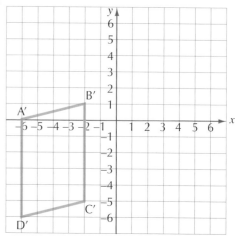

10B Planes of symmetry

1 Write down the number of planes of symmetry that each shape has.

a

3 cm

3 cm

b

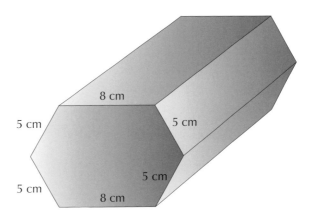

8 cm

5 cm

5 cm

5 cm

5 cm

5 cm

8 cm

2 Sketch a solid with exactly one plane of symmetry. Use dotted lines to show the plane of symmetry.

3 Copy each shape onto isometric paper. Draw a plane of symmetry using dotted lines or shading. Make a copy of the shape for every plane of symmetry.

a

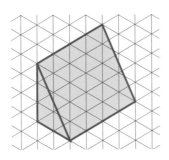

b

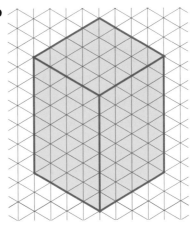

c

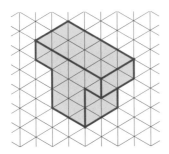

5

1 Write each of these map scales as a map ratio, i.e. 1 : number.

   a 1 cm to 3 km   **b** 1 cm to 100 m   **c** 2 cm to 100 km

 2 This map shows an area of England and Wales. Find the shortest (straight line) distance between these pairs of places. Give your answers to the nearest 5 km.

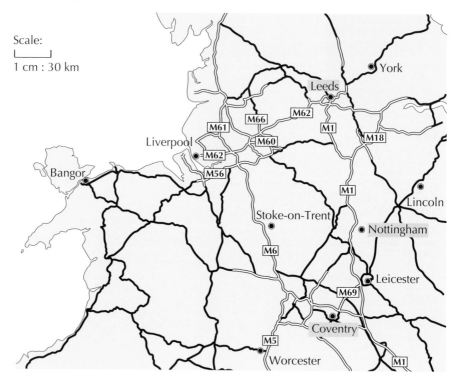

  a Liverpool and York   **b** Coventry and Bangor
  c Worcester and Leeds   **d** Lincoln and Stoke-on-Trent

3 The distance between Knutsford and Littleborough is 42 km.
How far apart would these towns be on the above map?

4 Barry made a journey using these motorways.
Nottingham to Leicester on the M1
Leicester to Coventry on the M69
Coventry to Stoke-on-Trent on the M6

Use a piece of string on the above map to estimate the total length of his journey, to the nearest 10 km.

5 The ratio of a map is 1 : 200 000.

  a Two places are 15 cm apart on the map. What is their actual distance apart?
  b Two places are 71 km apart. How far apart are they on the map?

10D Congruent triangles

Give reasons for your answers. State which condition of congruency you are using, i.e. SSS, SAS, ASA or RHS.

1 Show that these pairs of triangles are congruent.

a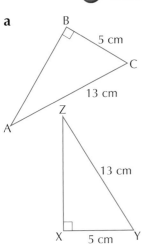

b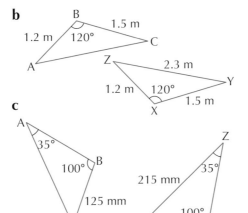

c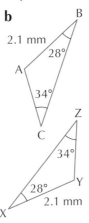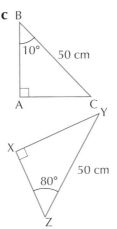

d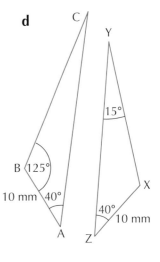

2 Which of these pairs of triangles are congruent?

a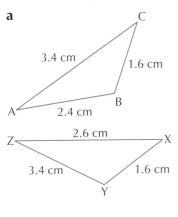

b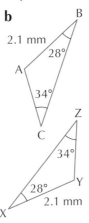

c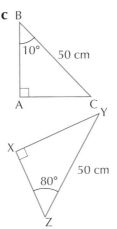

d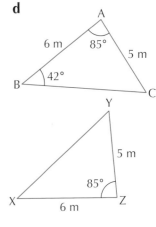

3 a Explain why these triangles are not necessarily congruent.

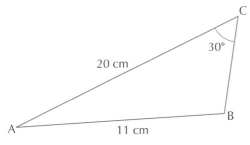 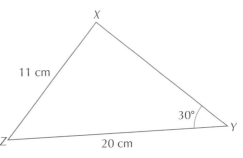

b Check your answer by trying to draw one of the triangles.

4 Draw a kite and label its vertices ABCD. Draw the diagonals and label their intersection E. State all pairs of congruent triangles.

CHAPTER 11 Algebra 5

11A Expansion

For Questions 1–3, expand the expressions.

1
a $5(d + 2)$
b $7(2p - 1)$
c $4(3 + m)$
d $10(5 - 3i)$
e $-(3a - 2)$
f $4(-2H + 6)$
g $-2(2w + 3)$
h $-6(5 - 6x)$

2
a $c(c + 7)$
b $m(5 - m)$
c $t(3t - 1)$
d $u(4 - 5u)$
e $d(e + f)$
f $-w(w - k)$
g $a(2a + b)$
h $h(3i - 4h)$

3
a $2m(m + 1)$
b $3u(4 - u)$
c $5p(2p + 3)$
d $-3x(5 - 2x)$
e $4i(j + 3i)$
f $3d(2e + 3f)$

For Questions 4 and 5, expand and simplify the expressions.

4
a $4(x + 5) - 2x$
b $5(2y + 1) + 2y - 3$
c $-3(t + 4) + 7t$
d $10 - 5(x - 1)$
e $9g - 2(g + 3)$
f $p - 4(2 - p)$

5
a $4(x + 3) + 2(x - 2)$
b $2(i - 1) + 3(2i + 1)$
c $5(2n - 3) + 3(4n + 1)$
d $4(r + 2) - 2(r - 3)$
e $3(2k + 3) - 2(k + 4)$
f $4(3j - 4) - 5(2j - 3)$
g $2(3 - 2f) + 5(4 + 3f)$
h $3(6 + u) - 4(3 - 5u)$

6 Make a sketch of this flag.

a Find an expression for the missing lengths and mark them on your diagram.
b Find expressions for:
   i the yellow area
   ii the white area.

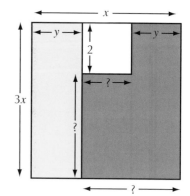

7
a Find an expression for the missing length.
b Find two different expressions for the shaded area.
c Show that your expressions are equivalent.

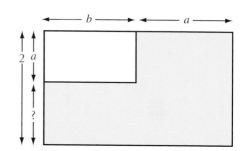

11B Factorisation

For Questions 1–4, factorise the expressions.

1 **a** $5m + 20$ **b** $12x - 18$ **c** $30 + 8k$ **d** $14 - 21h$

2 **a** $3m - 3n$ **b** $16p + 8q$ **c** $27a - 18b$ **d** $-20f - 10g$

3 **a** $ab + 2a$ **b** $mn - m$ **c** $2st + 5s$ **d** $de + df$
  e $4ij + 4i$ **f** $9r - 6rw$ **g** $15pq + 20pr$ **h** $6hu + 12iu$

4 **a** $d^2 + 3d$ **b** $4t - t^2$ **c** $9y^2 - 9y$
  d $8i + 6i^2$ **e** $p^2 - p$ **f** $2d + 3d^2$
  g $10n - 25n^2$ **h** $-7m^2 - 14m$

5 The tens digit of the number 63 is twice the units digit. 63 is divisible by 21.

  a Find more two-digit numbers with this property.
  b Find an expression for any two-digit number where the tens digit is twice the units digit.
  c Simplify your expression to show that such numbers are always divisible by 21.

Practice

11C Quadratic expansion

Expand and simplify these expressions.

1 **a** $(x + 3)(x + 1)$ **b** $(x + 9)(x + 2)$ **c** $(x + 5)(x + 4)$
  d $(x - 3)(x - 5)$ **e** $(x - 1)(x - 10)$ **f** $(x + 5)(x - 2)$
  g $(x - 7)(x + 4)$ **h** $(x - 6)(x - 4)$ **i** $(x + 4)^2$
  j $(x - 7)^2$ **k** $(x - 8)(x + 8)$

2 **a** $(m + n)(a + b)$ **b** $(2 + t)(v + w)$ **c** $(m + n)^2$
  d $(s + t)(s - t)$ **e** $(d - e)^2$

3 **a** $(2x + 3)(x + 1)$ **b** $(x + 5)(3x + 2)$ **c** $(4x + 5)(2x + 3)$
  d $(5x + 1)^2$ **e** $(4x - 1)(x - 3)$ **f** $(2x + 5)(x - 6)$
  g $(3x - 2)(2x + 3)$ **h** $(4x - 3)^2$

Practice

11D Substitution

1 **a** Make m the subject of the formula $L = m + n$.
  b Make h the subject of the formula $A = bh$.
  c Make d the subject of the formula $v = \frac{d}{t}$.
  d Make R the subject of the formula $D = x - R$.

2 Boxes contain 6 eggs. A crate contains m boxes of eggs. The number of eggs, N, contained in n crates is given by the formula $N = 6mn$.
Make m the subject of the formula.

3 The time, T minutes, to roast a joint of meat weighing W grams is given by the formula: $T = 20 + \dfrac{W}{25}$.

  a Calculate the time needed to roast a joint of meat weighing 800 g.
  b Make W the subject of the formula.
  c Calculate the weight of a joint that needs 1 hour 20 minutes of roasting time.

4 The total number of legs, L, in a classroom is given by the formula:

$$L = 4c + 2p$$

where c is the number of chairs and tables and p is the number of people.

  a Make p the subject of the formula.
  b Use your formula to calculate the number of people in a classroom containing 16 chairs and 6 tables given that there are 112 legs altogether.

5 **a** Explain why the area of this shape is given by the formula $A = 2ab - 9$.
  b Make a the subject of the formula.
  c Calculate a when $b = 8$ cm and the area is 135 cm².

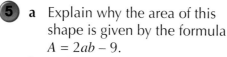

 6 An elastic string of natural length l metres is stretched by an amount x metres. The energy, E units, contained in the string is given by the formula:

$$E = \frac{2x^2}{l}$$

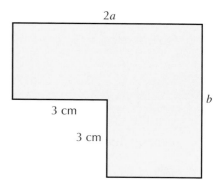

  a Calculate the energy contained in a string of natural length 200 cm when it is stretched 20 cm.
  b An elastic string has been stretched 30 cm and contains 5 units of energy. Calculate its natural length.
  c An elastic string of natural length 40 cm contains 2 units of energy. How far has it been stretched?

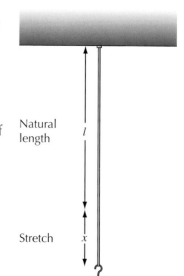

Practice

11E Graphs from equations in the form Ay ± Bx = C

1 Use a grid that goes from −10 to +10 on both *x* and *y* axes to draw the graphs of:

a $y - 4x = 8$ b $3y - 2x = 6$ c $x - 4y = 8$
d $2y + 5x = -10$ e $3y - 5x = 15$ f $4x - 3y = 12$
g $y + 4x = -8$ h $2x - 3y = 12$ i $y - 5x = 10$

CHAPTER 12 Soving problems and revision

Practice

12A Fractions, percentages and decimals

Do not use a calculator for Questions 1–4. Show your working.

1   a   What fraction of this shape is orange?
    b   What percentage is white?

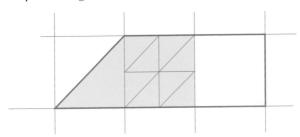

    c   i   What fraction of the whole shape is one small triangle?
        ii   Convert your answer to a decimal.

2   These three paint tins are identical. The fractions show how much paint each tin contains.

1/4 full of paint   2/5 full of paint   empty mixing tin

    a   Moira pours the paint from the first two tins into the mixing tin.
       What fraction of the mixing tin is filled with paint?
    b   She fills $\frac{3}{8}$ of the white tin with the paint mixture.
       What fraction of the mixing tin is now filled with paint?

c A full tin has a capacity of 600 ml. How much paint is now in the white tin?

d Altogether, Moira has $2\frac{3}{4}$ tins of white paint and $1\frac{5}{6}$ tins of blue paint. How many tinfuls of paint does she have altogether?

3 Tobo used 1.65 kg of sugar to make some plum jam and 750 g of sugar to make some apricot jam. Work out these using kilograms.

    a How much sugar did he use altogether?

    b He opened a new 4.15 kg bag of sugar to make the jam. How much sugar was left over?

    c Sugar costs 80p per kilogram. What was the cost of the sugar used for the jam?

    d Tobo made 5.76 kg of jam altogether. He filled 9 identical jars with the jam. How much jam did each jar contain?

4 450 spectators attended an annual cricket club match. Of these, 75 were women and 32% were club members.

    a How many club members attended?

    b What percentage of the spectators were women? Round your answer to the nearest whole number.

You may use a calculator for Questions 5–7.

5 **a** **i** How much VAT at 17.5% is paid on the MP3 player?

       ii What is the total cost of the MP3 player?

    b What is the cost of the headphones, after the discount?

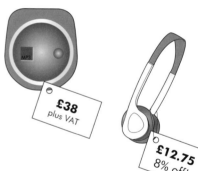

£38 plus VAT

£12.75 8% off!

6 The diameter of Mercury is 38% of the diameter of Earth. The diameter of the Earth is 12 756 km.

    a Round 38% and 12 756 km to 1 significant figure.

    b Use your approximations to estimate the diameter of Mercury.

7 This table shows the value of Tim's shares.

Share	Number of shares	Total value of shares
FirstBank	400	£3300
Pinwheel	50	£853
PowerCo	2000	£3720

a Which is the cheapest share?

b Pinwheel shares increased in value by $2\frac{1}{4}\%$.
What is the total value of Tim's Pinwheel shares?

c PowerCo shares decreased in value by 8.2%.
What is the value of a single PowerCo share?

8 Estimate answers to these.

a 0.38×63 b $17.8 \div 0.029$ c $\dfrac{41.7}{0.12 \times 7.9}$

Practice

12B Four rules, ratio and directed numbers

Do not use a calculator for Questions 1–4. Show your working.

1 a Calculate the following.
  i $18 - 9 \div 3$ ii $2 \times 3^2 - 1$ iii $\dfrac{2^3}{2^2 - 3}$

  b Insert brackets to make each equation true.
  i $24 \div 4 + 4 - 2 = 1$ ii $3 \times 4 + 2 \times 5 - 3 = 39$

2 Calculate these.

  a $-5 - 4$ b $2 - -3$ c $4 \times (-3)$
  d $-21 \div -7$ e $9 - 6 \div (-2)$ f $(-5 + 8)(-2 - 6)$

3 Terri is making cheesecakes.
Each cheesecake has a pastry
shell with filling.

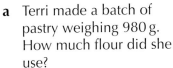

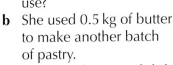

  a Terri made a batch of
  pastry weighing 980 g.
  How much flour did she
  use?

  b She used 0.5 kg of butter
  to make another batch
  of pastry.
  How much pastry did she make altogether?

  c Each cheesecake needs 370 g of pastry.
  How many cheesecakes could she make?

  d How much of each ingredient did she need to make all of the fillings?

4 Calculate these.

  a 32×0.7 b $18 \div 0.2$ c 0.3×0.02

Pastry recipe

Flour : Butter
= 5 : 2

Filling recipe
Makes 2 cheesecakes

Soft cheese   200 g
Sugar   75 g
2 eggs
Soured cream   125 ml

You may use a calculator for Questions 5–7.

5 Which perfume is cheaper?

6 The table shows details of some lorry journeys.

Journey	Distance travelled	Time taken	Average speed
Marston to Surfley	80 miles	$1\frac{1}{2}$ hours	
Deechurch to Creek	40 km		25 km/h
Penwood to Scotbridge		12 minutes	40 mph

Copy and complete the table.

7 **a** Tamsin converted £200 to Euros to buy some fruit juice. How many Euros did she receive?

Exchange rates
£1 = €1.48

€3.50 per bottle
€19 per case

b She bought as many bottles of fruit juice as possible. How much fruit juice did she buy?

c **i** How much money did she have left over?
   ii Convert your answer to British pounds.

Do not use a calculator. Show your working.

 a Copy and complete this sequence of calculations.

$3^2 - 1^2 =$ $4^2 - 2^2 =$
$5^2 - 3^2 =$ $6^2 - 4^2 =$

b Describe the sequence of answers to your calculations.
c Describe a rule to find the answer without squaring the numbers.
d Use your rule to find the answer to $15^2 - 13^2$.

2 Solve these equations.

    a $5d = -30$ **b** $\frac{p}{3} = 7$ **c** $4s - 3 = 17$

    d $9x = 7x + 12$ **e** $4m - 2 = m + 10$

3

FOR HIRE

£3 per day

£16 per week

    a Write down a formula for calculating the cost of hiring the cement mixer for w weeks and d days.
    b Use your formula to calculate the cost, £C, of hiring the cement mixer for
        i 3 weeks and 2 days   **ii** 90 days.
    c Martin pays £108 to hire the cement mixer for w weeks and 4 days.
        i Write down an equation.   **ii** Solve your equation to find w.

4 **a** Given that $a = 2$ and $b = -3$, evaluate these expressions.

        i $5a + b$ **ii** b^2 **iii** $3(a - b)$

    b Expand and simplify these expressions.
        i $2(x + 4)$ **ii** $5(p - 1) - 3p$
        iii $12(t + 1) + 3(t + 2)$
        iv $5(m + 2) - 3(m - 1)$
    c Factorise these expressions.
        i $16d + 10$ **ii** $t^2 - t$ **iii** $9f - 4fg$
        iv $12ab + 8a$ **v** $8y^2 + 10y$

5 The diagram shows a betting game. To play the game costs x pence. A counter is tossed on to the board. The winnings are shown on the board.

Bet x pence

| A WIN $x + 6$ | B LOSE $x + 4$ |
| C WIN $2x - 4$ | D LOSE $3x$ |

a Paddy flips a counter on to area A five times in a row. Write an expression to show how much he won altogether. Simplify your expression.

b He then flips a counter on to area B twice, C once and D once. Write an expression to show his total winnings. Simplify your expression.

c Paddy won 32p altogether. Write an equation involving x. Solve your equation to find the amount it costs to play the game.

d Do you think the game is fair? Justify your answer using algebra.

Practice

12D Graphs

Do not use a calculator. Show your working.

1 a Write down the equation of line **i** A **ii** B **iii** C.

b Write down the equation of the x-axis.

c i Copy the grid and draw the line with equation $y = 2x - 2$.

ii Where does this line intersect line C?

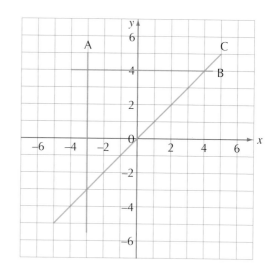

2 a i   Copy the diagram.
      ii Add point D to the
          diagram so that ABCD
          is a parallelogram.
      iii Write down the
          coordinates of D.

   b Which vertices of the
      parallelogram lie on each
      of these lines?
      i $x = -4$
      ii $y = -x$
      iii $y = x + 2$

   c i   Write down the
          gradient of line AD.
      ii Extend AD to intersect
          the y-axis. Write down
          the y-intercept.
      iii Write down the equation of line AD in the form $y = mx + c$.

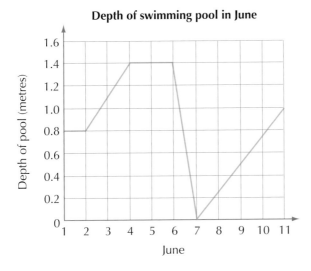

3   The normal depth of a pool is 1 metre. The graph shows the depth of water
    in an outdoor swimming pool at the end of each of the first 11 days of June.

Depth of swimming pool in June

   a On which day was the pool emptied?
   b It rained on two consecutive days. When did it rain?
   c i   How long did it take to refill the pool to the normal depth?
      ii By how much did the depth increase per day?
   d When was the rate of change in the depth of the pool the greatest?
      Give a reason for your answer.

4 The chart shows the graphs of legal gun ownership and crimes involving firearms from 1979 to 1992.

Legal gun ownership and robberies involving firearms
Source: http://members.aol.com/gunbancon/gifs/robbery.gif

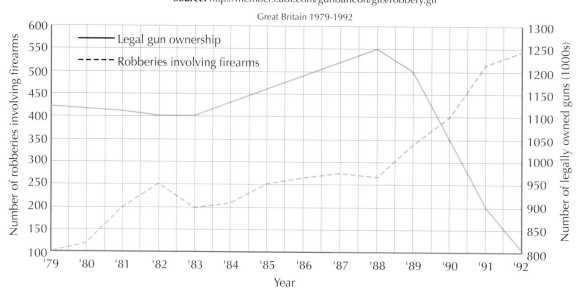

a What is the difference in legal gun ownership between 1983 and 1988?

b When was the fastest increase in crimes involving firearms?
Explain your answer.

c Is this a valid conclusion?
'During the period 1983 to 1988, legal gun ownership increased and so did crimes involving firearms. This shows that legal gun ownership leads to an increase in gun crime.'
Give evidence to support your answer.

5 Here are the equations of eight graphs.

A $x = 4$

B $y = 4x - 4$

C $y = \frac{1}{4}x$

D $y = -4$

E $y = 4x$

F $y = -4x$

G $y = -\frac{1}{4}x + 4$

H $y = 4$

a Which pairs of graphs are parallel?
b Which pairs of graphs are perpendicular?
c Which graphs pass through the origin?
d Which graphs have a y-intercept of 4?
e Which graphs cross the x-axis at 1?
f Which graphs pass through the point $(1, -4)$?

Practice

12E Geometry and measures

You may use a calculator. Show your working.

(1) Copy these shapes.

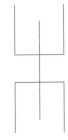

A

B

C

D

a Use dotted lines to show any lines of symmetry.
b Write down the degree of rotational symmetry below each shape. Mark the centre of rotation using a dot.
c Copy shape C again. Add more lines to give the shape exactly two lines of symmetry.

(2)

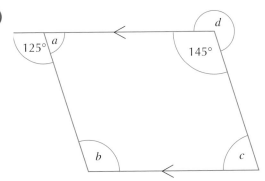

125° *a*

d

145°

b

c

a Which of the marked angles are:
  i obtuse?   ii reflex?   iii acute?
b Calculate the marked angles.

(3) a Construct this triangle accurately.
b Measure the side AC.
c What type of triangle is this?
d Construct the perpendicular bisector of angle ACB.
e i Measure the height of the triangle.
  ii Calculate the area of triangle ABC.

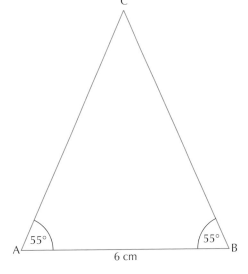

C

55°   55°
A   B
          6 cm

4 This diagram shows two regular hexagons.

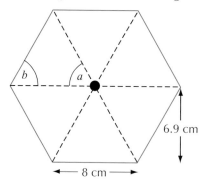

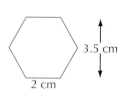

a   Calculate the marked angles.
b   Calculate the area of one of the small triangles.
c   Calculate the area of the large hexagon.
d   Calculate the area of the small hexagon.
e   Do regular hexagons tessellate? Illustrate your answer with a diagram.

5 This diagram shows a brick and a patio made using such bricks.

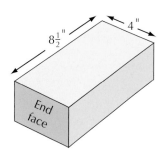

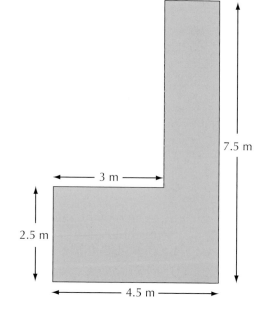

a   Given that $1'' \approx 28$ mm, calculate the area of the top of the brick, in cm^2.
b   The area of the end face of the brick is 95 cm^2.
    Calculate the height of the brick, to the nearest millimetre.
c   Calculate the volume of the brick.
d   Calculate the area of the patio.
e   Estimate the number of bricks needed to cover the patio area.

6

a Describe the transformation that maps shape P on to shape Q.

b Copy the diagram, without shape Q. Extend the x- and y-axes to +12 and –12.

c Reflect shape P in the x-axis. Label the image R.

d Enlarge shape P using centre X and a scale factor of 3. Label the image S.

e Rotate shape P 90° clockwise about the centre (–4, 3). Label the image T.

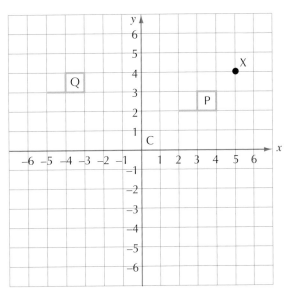

7

a Calculate the length, *d*, of rope connecting the sailboarder to the kite.

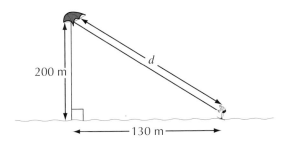

b Calculate the width, *x*, of the cellar opening.

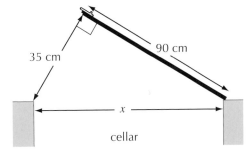

Practice

12F Statistics

You may use a calculator.
Show your working.

1 A fruit machine has three wheels. Each wheel has 25 symbols. The table shows the number of each symbol on the first wheel.

Symbols on wheel 1	
Apple	7
Cherry	2
Banana	4
Pear	5
Bell	1
Orange	6

a For each pull of the handle, what is the probability that the first wheel will land on:
  i a cherry? **ii** a fruit? **iii** an apple or orange?
b Which of the events in part **a** are unlikely to occur?
c The probability of the second wheel landing on an orange is 0.2.
  i What is the probability of it not landing on an orange?
  ii How many oranges would you expect this wheel to land on with 150 pulls of the handle?
  iii How many oranges are on the second wheel?
d In 20 pulls of the handle, the third wheel landed on a bell 3 times.
  i What is the experimental probability of the third wheel landing on a bell?
  ii Estimate the number of bells on the wheel.
  iii How could you obtain a better estimate?

2 StarGirl is a handheld electronic game with five levels of difficulty. When a player scores 100 points in one turn, they start the next level.

a Marcus obtained these scores on Level 1.

20 0 60 40 80 40 40 80 60 0 100

Calculate these statistics.
  i Mode **ii** Median **iii** Mean **iv** Range
b At the end of Level 1, Marcus received a bonus of 60 points. Explain how this affects each of the statistics in part **a**.
c This table summarises Marcus' scores until he completed Level 5.

Score	Number of turns
0	12
20	20
40	45
60	38
80	29
100	5

  i What is his modal score?
  ii Calculate the total points he scored.
  iii Calculate his mean score.

3 This diagram shows a balanced spinner and an ordinary dice.

The spinner is spun and the dice is rolled at the same time. The numbers they land on are added together.

a Copy and complete this table of possible outcomes.

		Dice					
		1	**2**	**3**	**4**	**5**	**6**
Spinner	**1**	2					
	1						
	2						
	3						
	3						

b Which totals are the least likely?

c Calculate the probability of obtaining
  i a total of 4.   **ii** an even total.
  iii a total of less than 6.   **iv** a total that is a prime number.

4 Hing Wai conducted an investigation into the amount of fiction and non-fiction adults watch on TV. He asked ten of his teachers these questions.

'How many hours do you spend watching films each week?'
'How many hours do you spend watching documentaries each week?'
His report included this table of results and scatter diagram.

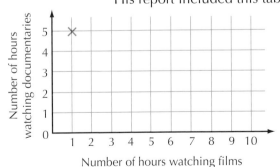

Number of hours watching documentaries

Number of hours watching films

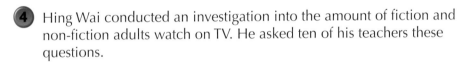

Teacher	Ms Witon	Mr Carter	Mr Singh	Mrs Tenby	Mrs Dean	Mr Chan	Ms Aldridge	Mr Jones	Ms Foster	Mr Phillips
Number of hours watching films	0	1	5	4	3	2	5	3	4	3
Number of hours watching documentaries	4	5	0	2	4	3	3	4	3	2

a Copy and complete the scatter diagram.
b Draw a line of best fit on your diagram.
c Describe any correlation.
d Use your line of best fit to estimate
    i the number of hours spent watching films when $3\frac{1}{2}$ hours are spent watching documentaries
    ii the number of hours spent watching documentaries when 9 hours are spent watching films.
e What conclusion could Hing Wai make, based on this data?
f Give a reason why this conclusion may not be true.
g Write down another useful question that Hing Wai could have asked.

CHAPTER 13 Statistics 3

Practice

13A Statistical techniques

1 Calculate the mode, the median and the mean for each set of data below.

    a 3, 8, 9, 13, 14, 15, 15
    b 1, 2, 2, 2, 4, 10, 10, 11, 12
    c £1.50, £3.50, £1, £2.50, £1.50
    d 12, 12, 13, 15, 18, 20

2 Criticise each of the following questions that were used in a questionnaire about pupils' music habits.

    a How do you listen to your favourite music?
        ☐ MP3 player ☐ CD player   ☐ Computer
    b How long do you spend listening in the evenings?
        ☐ 0–1 hour   ☐ 1–2 hours   ☐ 2–3 hours   ☐ 3–4 hours
    c How many tracks of music do you have?
        ☐ 0–500   ☐ 500–1000   ☐ 1000–1500   ☐ other

3 A football team uses 21 players through the season. The number of matches each player has played in is given below:

11, 9, 21, 19, 18, 18, 19, 22, 17, 19, 7, 3, 6, 11, 9, 13, 11, 18, 17, 11, 6

    a Use the data to copy and complete the stem and leaf diagram:
        0
        1
        2
        key: | means ☐
    b Work out the median mark.
    c State the range of the marks.
    d How many players played more than the mode?

4 A pub quiz team is made up of people of different ages.
The table shows the number of people in each age group.

Age group	Number
18 to 22	2
23–30	1
31–50	2
over 50	3

   a Represent this information in a pie chart.
   b Amy says, "This shows that the same percentage of under 23s were
      chosen as 31 to 50s." Explain why this statement might not be true.

5 Below are the scores of a class of 30 in a maths exam:

Boys	56, 17, 28, 87, 61, 90, 45, 37, 35, 48, 52, 58, 35, 9, 11, 78, 82
Girls	18, 93, 56, 75, 82, 55, 43, 49, 52, 28, 50, 44, 81

   a Copy and complete the two-way table to show the frequencies:

	Boys	Girls
$0 \leq M \leq 20$		
$20 < M \leq 40$		
$40 < M \leq 60$		
$60 < M \leq 80$		
$80 < M \leq 100$		

   b What percentage scored more than 60%?
   c What is the modal group for the girls?
   d In which group is the median score for the boys?

Practice

13B A handling data project

Investigate one of the following topics:

1 Compare the number of pages in books with the number of chapters.

2 Compare the number of digits that can be accurately remembered by
different aged people.

3 Investigate how accurately different people can estimate the weight of
household objects. Compare different ages or male and female.

4 Choose an investigation of your own.

CHAPTER 14 Geometry and Measures 4

5

1 Find
  i the perimeter and
  ii the area of each of the following rectangles:

a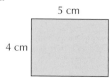
5 cm
4 cm

b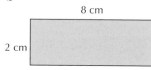
8 cm
2 cm

c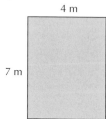
4 m
7 m

6

2 Find the area of each of the following shapes:

a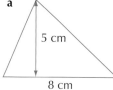
5 cm
8 cm

b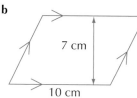
7 cm
10 cm

c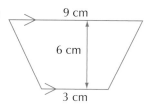
9 cm
6 cm
3 cm

3 Calculate
  i the circumference and
  ii the area of each of the following circles:

Take π = 3.14 or use the π key on your calculator. Give your answers to one decimal place.

a
4 cm

b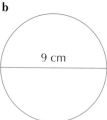
9 cm

c
4.2 cm

4 Calculate
  i the surface area and
  ii the volume of each of the following cuboids:

a

b

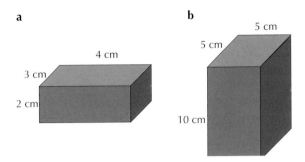

5 Calculate
  i the surface area and
  ii the volume of each of the following 3-D shapes:

a   **b**

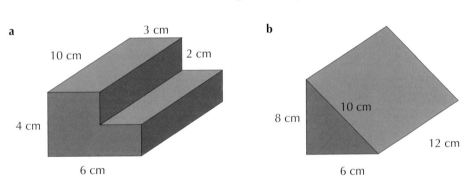

Tiling patterns

Here are two square tiles, made
from red squares and blue squares.

This 3 by 3 tile is made from 1 blue square
and 8 red squares.

This 4 by 4 tile is made from 4 blue square
and 12 red squares.

6

① Draw diagrams to show how many blue squares and red squares are needed for 5 by 5 and 6 by 6 tiles.

② Draw a table to show your results and write down any patterns you notice.

③ Can you predict how many blue squares and red squares are needed for a 7 by 7 tile?

④ How many blue squares and red squares are needed for an n by n tile?

Practice

14C Symmetry revision

① Copy each of these shapes and draw its lines of symmetry. Write below each shape the number of lines of symmetry it has.

a 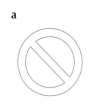   b 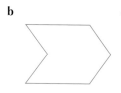   c   d   e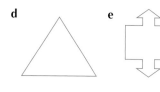

② Copy each of these shapes and write the order of rotational symmetry below each one.

a    b 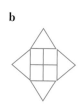   c 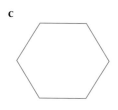   d   e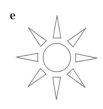

③ Write down the number of planes of symmetry for each of the following 3-D shapes.

a   b    c   d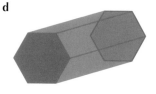

Cube   Cuboid   Square-based pyramid   Regular hexagonal prism

5

7

14D Symmetry investigations

Symmetry in regular shapes

Here is a regular hexagon with all its diagonals drawn in.

1 **a** How many different symmetrical shapes can you find inside the regular hexagon?

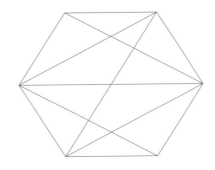

b On copies of the diagram, colour them in to make shapes which have either line symmetry or rotational symmetry.
   i Use two different colours at first.
   ii Then try using three different colours.

2 Now repeat the investigation by using a regular pentagon.

Write down anything you notice.

CHAPTER 15 Statistics **4**

15A Revision of probability

1 Two dice are rolled together.

   a How many different outcomes are there? Create a sample space showing all the possibilities.
   b What is the probability of getting a double?
   c What is the probability of getting exactly one six?
   d What is the probability of getting no sixes at all?

2 Joseph has difficulty waking up on time:

   The probability that he wakes up on time is 0.2.
   The probability that he wakes up late is 0.7.
   a What is the probability that Joseph wakes up early?
   b What is the probability that he does not wake up late?

3 A scientist asks fifty 14-year-olds how many of them can remember a seven-digit number. He finds that eight of them can remember all seven digits, while 15 of them can remember six of the digits.

Use these results to estimate the probability that a 14-year-old chosen at random has:
a remembered all seven digits.
b remembered six of the digits.
c cannot remember six or seven digits.

4 An eight-faced spinner is spun 100 times. Here are the results:

Number on spinner	1	2	3	4	5	6	7	8
Frequency	13	16	8	7	17	15	8	16

a What is the experimental probability of the spinner landing on the number 6?
b Write down the theoretical probability of a fair, eight sided spinner landing on the number 6.
c Consider whether you think the spinner is fair.
d How many sixes would you expect from this spinner if it was spun 250 times?

Practice

15B A probability investigation

Carry out an experiment to investigate one of the following:

1 When you roll two dice, you get a total of 7 more times than any other total.

2 If you shuffle a pack of cards and deal out the top two cards, half the time these two cards will both be the same colour.

3 If you flip three coins, how often will you will have at least two heads?

4 With a newspaper or magazine open in front of you, shut your eyes and, with a pencil, point it anywhere on the pages and see how often it actually lands on words, pictures or a space.

5 Choose an investigation of your own.

 **16** GCSE Preparation

Practice **16A Multiplication**

Do not use a calculator. Show your working.

1 Calculate these.

a 9×443	**b** 722×5	**c** 8×3283
d 7104×6	**e** 44×87	**f** 93×26
g 13×564	**h** 756×55	**i** 39×604
j 628×88	**k** 99×482	**l** 507×47
m $22 \times 11 \times 76$	**n** $53 \times 8 \times 87$	

2 A full matchbox contains 36 matches. A pack contains 8 matchboxes. How many matches are contained in

a a pack? **b** 75 matchboxes? **c** 24 packs?

3 Shania works 28 hours a week for £7.31 per hour. Sui Main works 43 hours a week for £4.62 per hour.

a How much does each person earn per week? Work in pence.
b What is the difference in their weekly earnings? Work in pence.

4 Warren types 54 words per minute. How many words does he type in

a 19 minutes? **b** 48 minutes?
c 2 hours? **d** 4 hours 17 minutes?

5 The north and south sides of a stadium each have 35 rows of 126 seats. The east and west sides each have 18 rows of 199 seats. How many seats are there altogether?

Practice **16B Division**

Do not use a calculator. Show your working.

1 Calculate these. Some of the calculations have a remainder.

a $513 \div 9$	**b** $\frac{857}{4}$	**c** $7288 \div 8$
d $\frac{9040}{6}$	**e** $627 \div 11$	**f** $429 \div 13$
g $\frac{836}{20}$	**h** $588 \div 21$	**i** $826 \div 35$
j $\frac{999}{68}$	**k** $693 \div 24$	**l** $972 \div 18$

2 **a** How many 9 cm chains can be cut from a length of 283 cm?
   b How much chain is left over?

3 12 people share a prize of £859. How much does each person receive, to the nearest pound?

4 Which purchase is the best value? Work in pence.

5 Calculate the width of this playing field.

Width
Area = 896 m²
62 m

Practice

16C Problems involving multiplication and division

Do not use a calculator. Show your working.

1 Maths Frameworking pupil books are packed into boxes of 18. Delivery vans can carry 424 boxes at a time.

   a How many books can a delivery van carry?
   b How many boxes are needed to pack 702 books?

2 These shapes all have the same area.

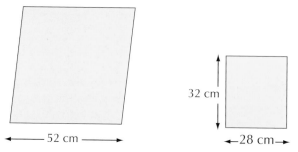

52 cm
32 cm
28 cm
11 cm

   a Calculate the area of the rectangle.
   b Calculate the height of the parallelogram.
   c Calculate the base of the triangle.

3 Hotel rooms cost £23 per person per night.

  a How much will it cost for a single person to stay for 14 nights?
  b How much will it cost for three people to stay for 9 nights?
  c Samuel's hotel bill was £552. How many nights did he stay?
  d The hotel was fully booked for three days and took £828 in rent.
      How many people stayed at the hotel each night?

4 Mariam downloaded 928 kB of data from the Internet in 18 seconds.

  a How much data did she download each second, to the nearest kB?
  b How much data could she download at the same rate in:
      i 45 seconds?   **ii** 7 minutes?

5 Quentin used 629 cl of paint to colour 37 m² of floor. Each metre of floor is made from 16 square tiles. How much paint is needed to cover:

  a one square metre of floor?
  b a floor of area 930 m²?
  c 448 tiles?

Practice

16D Fractions

Do not use a calculator unless instructed.

1 Cancel these fractions.

  a $\frac{10}{15}$ **b** $\frac{9}{21}$ **c** $\frac{16}{36}$ **d** $\frac{24}{40}$ **e** $\frac{18}{8}$

2 Fill in the missing numbers.

  a $\frac{18}{30} = \frac{\square}{15} = \frac{6}{\square}$ **b** $\frac{12}{20} = \frac{60}{\square} = \frac{\square}{5}$

  c $\frac{16}{24} = \frac{\square}{30} = \frac{24}{\square}$ **d** $\frac{49}{28} = \frac{14}{\square} = \frac{\square}{20}$

3 Calculate these.

  a $\frac{3}{5}$ of £80   **b** $\frac{4}{7}$ of 63 kg

  c $\frac{7}{8}$ of 120 cm   **d** $\frac{3}{4}$ of 3000 cl

4 Order each pair of fractions using the < or > sign.

  a $\frac{3}{7}, \frac{4}{9}$ **b** $\frac{5}{8}, \frac{7}{12}$ **c** $\frac{2}{3}, \frac{7}{11}$ **d** $\frac{2}{9}, \frac{5}{21}$

5 Arrange these from smallest to largest.

  $\frac{7}{10}$ of 110, $\frac{2}{7}$ of 287, $\frac{9}{11}$ of 77, $\frac{5}{9}$ of 144

6 A measuring jug contains 600 ml when full.

  a **i** How much does it contain when $\frac{7}{12}$ full?
      ii What fraction is empty?
  b It takes $3\frac{3}{5}$ jugs to fill a saucepan.
      What is the capacity of the saucepan?

7 Jan threw a dart at a dartboard 84 times. $\frac{3}{7}$ of the darts hit the inner ring, $\frac{1}{6}$ hit the outer ring, $\frac{3}{14}$ hit the bull, and the remainder hit doubles or trebles.

    a How many darts hit
        i the inner ring?   **ii** the outer ring?
        iii the bull?   **iv** a double or treble?
    b What fraction hit a double or treble?

8 Which of these fractions is closest to $\frac{7}{10}$?

$$\frac{2}{3}, \frac{7}{11}, \frac{13}{20}, \frac{12}{17}$$

9 In 1990, the world released 21 600 million tons of carbon dioxide into the atmosphere for creating energy. This table shows the fraction some areas contributed. Copy and complete the table.

Country	Fraction	Carbon dioxide (millions of tons)
Europe	$\frac{3}{20}$	
North America		5400
Japan	$\frac{2}{45}$	
Rest of the world	$\frac{5}{9}$	

Practice

16E Adding and subtracting fractions

Do not use a calculator. Show your working.
Cancel answers and express as mixed numbers where necessary.

1 A bag of rice is $\frac{2}{3}$ full. An identical bag is $\frac{3}{4}$ full.

    a How many full bags of rice are there altogether?
    b If a full bag contains 720 g of rice, how much rice is contained in the two bags?
    c $\frac{5}{8}$ of a bagful of rice is used for a meal. What fraction of a bagful of rice is left?

2 Of the water consumed by a household each day, $\frac{3}{5}$ is used for showers, $\frac{4}{15}$ for cleaning and the rest for cooking and drinking.

    a What fraction of water is used for cooking and drinking?
    b 12 litres is used for cooking and drinking. How much water is used for showers?

3 Calculate these.

    a $\frac{3}{4} + \frac{1}{7}$ **b** $\frac{1}{6} + \frac{3}{8}$ **c** $\frac{7}{10} + \frac{4}{5}$

    d $\frac{2}{3} - \frac{1}{5}$ **e** $\frac{7}{8} - \frac{5}{6}$ **f** $\frac{1}{4} + \frac{1}{6} + \frac{5}{12}$

4 $\frac{1}{6}$ of a magazine is used for adverts, $\frac{2}{9}$ is devoted to a feature, and the rest is used for regular articles.

    a What fraction is used for regular articles?
    b The magazine has 54 pages. How many are devoted to the feature?

5 Marcel has tiled $\frac{5}{9}$ of his floor using 80 tiles.

    a What fraction of the floor is not tiled?
    b How many tiles are needed to cover the whole floor?

6 Calculate these.

    a $\frac{9}{10} - \frac{3}{8} - \frac{2}{5}$ **b** $2\frac{3}{5} + 1\frac{1}{2}$ **c** $3\frac{2}{3} - \frac{5}{9}$

    d $2\frac{1}{4} - 1\frac{5}{6}$ **e** $6\frac{1}{4} + 8\frac{2}{3}$ **f** $10\frac{5}{16} - 5\frac{1}{8}$

    g $8\frac{1}{6} - 6\frac{11}{12}$

7 Debbie measured these pipes using an old ruler.

A $2\frac{9}{10}$ ″

B $2\frac{3}{16}$ ″

C $3\frac{7}{12}$ ″

    a What is the total length of pipes A and B?
    b How much longer is pipe C than pipe B?
    c Which two pipes have the closest length?

8 Daniel uses $7\frac{5}{8}$ rolls of wallpaper for a bedroom and $4\frac{5}{6}$ rolls for the toilet.

    a How much wallpaper did he use altogether?
    b How much more wallpaper did he use for the bedroom than the toilet?

Practice

16F Multiplying and dividing fractions

Do not use a calculator. Show your working.
Cancel answers and express as mixed numbers where necessary.

1 Calculate these.

    a $\frac{4}{7} \times \frac{2}{3}$ **b** $\frac{3}{8} \times \frac{4}{5}$ **c** $\frac{9}{10} \times \frac{5}{6}$

    d $2\frac{2}{3} \times 1\frac{3}{4}$ **e** $2\frac{2}{9} \times \frac{12}{25}$ **f** $\frac{3}{4} \div \frac{4}{5}$

    g $\frac{9}{10} \div \frac{7}{8}$ **h** $\frac{15}{16} \div \frac{5}{8}$ **i** $1\frac{5}{6} \div 3\frac{1}{2}$

    j $\frac{7}{8} \div 1\frac{1}{6}$ **k** $\dfrac{2\frac{5}{8}}{2\frac{11}{12}}$

2 Calculate these.

    a $\frac{2}{3} \times \frac{5}{8} - \frac{1}{4}$ **b** $(1\frac{4}{5})^2$ **c** $\frac{5}{8}(1\frac{2}{3} + 3\frac{1}{2})$

3 **a** Calculate the perimeter and area of this parallelogram.

$1\frac{5}{16}"$ $5\frac{1}{4}"$ $1\frac{1}{8}"$

b Calculate the length of the base of this rectangle.

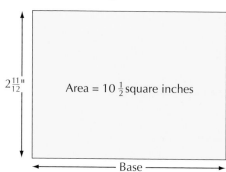

$2\frac{11}{12}"$ Area = $10\frac{1}{2}$ square inches   Base

4 A wheelbarrow can hold $\frac{2}{15}$ m³ of sand.

   a How much sand can be moved using 12 wheelbarrow loads?
   b How many wheelbarrow loads are needed to move $1\frac{5}{9}$ m³ of sand?

5 **a** Maria has drunk $\frac{2}{5}$ of a can of orange. What fraction of a can does she have left?
   b She drinks $\frac{3}{4}$ of the remaining orange. What fraction of a can did she drink?
   c What fraction of a can did she drink altogether?

6 A reel contains $3\frac{3}{8}$ metres of flex. Jane buys $1\frac{5}{8}$ metres. What fraction of the reel did she buy?

Practice

16G Directed numbers

Do not use a calculator. Show your working.

1 Calculate these.

a $-4 + 9$		**b** $-3 - 3$		**c** $10 - 18$	
d $6 - -4$		**e** $-8 - +3$		**f** $(-4) \times (-7)$	
g $2(-6)$		**h** $(-5)^2$		**i** $-4(-10)$	
j $20 \div (-5)$		**k** $\frac{-8}{-2}$		**l** $-2 - 3 \times (-4)$	
m $(-8 + 4)^2$		**n** $-6(-1 - 1)$		**o** $12 \div (-3) \times (-2)$	
p $\frac{-8 + 2}{-3}$		**q** $4 - (-2)^3$		**r** $\frac{(-3)(-8)}{(-2)^2}$	
s $\frac{-2 + 6(-2)}{-4 - +3}$					

2 Find the missing numbers.

a $-5 - \square = 9$

b $\square \times (-3) = 18$

c $-2(\square + 7) = 4$

d $4 - \square \div (-2) = 20$

e $\dfrac{-4 - \square}{3 - -2} = 3$

f $\dfrac{2(5 - -7)}{-2 + \square} = -8$

3 Given that $x = -3$, $y = 4$ and $t = -8$, evaluate these expressions.

a $-x - y$

b $5t - 2y$

c $\dfrac{t}{y}$

d $x(y + t)$

e $t^2 - y^2$

f $2xyt$

g $(3y - t)(2t + x)$

h $3y - 4x + 3t$

i $\dfrac{t - y}{2x}$

j $\dfrac{y}{-2} - \dfrac{12}{x}$

k $3(x + 2y) - 2(t - 3x)$

4 Use the cards to make calculations that give these answers. Each number card may only be used once.

| -6 | 3 | 8 | -2 | 2 | $($ | $)$ | $+$ | $-$ | $\div$ | $\times$ |

a -3

b 10

c 4

d 3

e -24

f 25

g 2

h -1

 Practice

16H Percentages

Do not use a calculator for Questions 1–6.

1 Calculate these.

a 5% of \$30

b 45% of 120 kg

c 95% of 3000 litres

d 15% of £26

e 400% of 13 tonnes

f 120% of 60

2 Increase each quantity by the given percentage.

a 90 m by 20%

b 240 g by 5%

c 16p by 30%

d 44 seconds by 15%

e £19 by 1%

f 30 minutes by 300%

3 Decrease each quantity by the given percentage.

a \$24 by 10%

b 5000 kg by 25%

c 80 m² by 35%

d 370 cl by 5%

e £6 by 2%

4 **a** Express 17 m as a percentage of 25 m.
b What percentage of 50 kg is 43 kg?
c What is £7 as a percentage of £20?
d Convert to a percentage.

i $\dfrac{78}{200}$

ii 0.06

iii $\dfrac{2}{3}$

iv $2\frac{1}{4}$

v 1.48

vi $\dfrac{13}{40}$

vii 9

5 2% interest was added to Mario's credit card balance of £740. What was his new balance?

6 A mail-order coat costs £190 after a reduction of 5%. What was the price of the coat before the reduction?

Use a calculator to answer the remaining questions. Round your answers to a suitable degree of accuracy, where necessary.

7 **a** Increase €63 by:
   i 22%   **ii** $17\frac{1}{2}$%   **iii** 2.5%   **iv** 129%
b Decrease £7.45 by:
   i 6%   **ii** 94%   **iii** $33\frac{1}{3}$%   **iv** 0.8%

8 A loaf of bread increased its height by 24% to 25 cm during baking. What was its height before baking?

9 A swimming pool contains 13 450 litres of water.

a A pump increases its volume by 7%.
   How much water does it now contain?
b 13% of the water was then drained away.
   How much water does it now contain?
c Every week, the pool loses 1.4% of its water due to evaporation.
   How much water does it contain
   i after 1 week?   **ii** after 4 weeks?

10 **a** What is the cost of the jacket including VAT?
b What was the cost of the trousers before VAT was added?

£94 excluding VAT @ 17.5%

£58 including VAT @ 17.5%

Practice

16I Algebra

1 Simplify these expressions.

a $3x - 5 + 6x + 6$	**b** $a + 4 - 3a - 7$	**c** $3y - -2y$
d $3d + 4e - 3e + 2d$	**e** $2 \times (-4t)$	**f** $5x \times 4x$
g $3ab - 2a + 4ab - 5a$	**h** $4y^2 - 2y + 6y^2 + 5y$	

2 Expand and simplify these expressions.

a $3(2x - 1)$	**b** $-(4y + 7)$	**c** $x(3 + 2x)$
d $-4(2 - 5y)$	**e** $4 + 2(3a + 2)$	**f** $6x - 3(x - 5)$
g $2d(e + 3) - d$	**h** $5p^2 - 3p(p - q)$	

3 Solve these equations.

   a $5t - 2 = 13$ **b** $\frac{x}{4} + 3 = 9$

   c $7d = 4d + 21$ **d** $2m - 4 = m - 6$

   e $-8w = 7$ **f** $2p = 9p - 35$

   g $3(x - 3) = 18$ **h** $-2(3y + 4) = 16$

4 Expand and simplify these expressions.

   a $5(x + 4) + 2(x - 1)$ **b** $9(y + 2) - 3(y + 1)$

   c $2(3e - 2) + 4(2e - 1)$ **d** $4(3d + 4) - 5(4d - 2)$

   e $2(3 - 2u) + 4(1 - 3u)$ **f** $x(5x + 2y) - 2x(x - 2y)$

5 Solve these equations.

   a $2(x + 3) + 3(x - 1) = 23$ **b** $2(2t - 1) + 3(t + 8) = 1$

   c $4(2y + 3) - 2(3y - 1) = 18$ **d** $2(p + 3) - 4(p - 1) = 7$

   e $9(m + 2) = 2(3m + 8) + 11$

6 **a**   Write an equation involving x. Solve your equation.

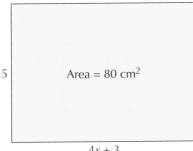

5   Area = 80 cm²

4x + 3

   b   Write an equation involving x. Solve your equation. Find the percentage of people who chose football as their favourite sport.

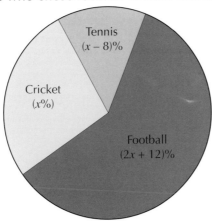

Tennis
$(x - 8)\%$

Cricket
$(x\%)$

Football
$(2x + 12)\%$

William Collins' dream of knowledge for all began with the publication of his first book in 1819. A self-educated mill worker, he not only enriched millions of lives, but also founded a flourishing publishing house. Today, staying true to this spirit, Collins books are packed with inspiration, innovation and practical expertise. They place you at the centre of a world of possibility and give you exactly what you need to explore it.

Collins. Freedom to teach.

Published by Collins

An imprint of HarperCollins*Publishers*

77–85 Fulham Palace Road

Hammersmith

London

W6 8JB

Browse the complete Collins catalogue at
www.collinseducation.com

10 9 8 7 6 5 4 3

ISBN 978-0-00-726806-1

Keith Gordon, Kevin Evans, Brian Speed and Trevor Senior assert their moral rights to be identified as the authors of this work.

British Library Cataloguing in Publication Data

A Catalogue record for this publication is available from the British Library.

Commissioned by Melanie Hoffman and Katie Sergeant

Project management by Priya Govindan

Covers management by Laura Deacon

Edited by Brian Ashbury

Proofread by Amanda Dickson and Tessa Akerman

Design and typesetting by Newgen Imaging

Design concept by Jordan Publishing Design

Covers by Oculus Design and Communications

Illustrations by Tony Wilkins and Newgen Imaging

Printed and bound by Printing Express, Hong Kong

Production by Simon Moore

Every effort has been made to trace copyright holders and to obtain their permission for the use of copyright material. The authors and publishers will gladly receive any information enabling them to rectify any error or omission in subsequent editions.

Mixed Sources

Product group from well-managed forests and other controlled sources
www.fsc.org Cert no. SW-COC-1806
© 1996 Forest Stewardship Council

FSC is a non-profit international organisation established to promote the responsible management of the world's forests. Products carrying the FSC label are independently certified to assure consumers that they come from forests that are managed to meet the social, economic and ecological needs of present and future generations.

Find out more about HarperCollins and the environment at
www.harpercollins.co.uk/green